# Atmospheres and Shared Emotions

This book explores the role atmospheres play in shared emotion. With insights from leading scholars in the field, *Atmospheres and Shared Emotions* investigates key issues such as the relation between atmospheres and moods, how atmospheres define psychopathological conditions such as anxiety and schizophrenia, what role atmospheres play in producing shared aesthetic experiences, and the significance of atmospheres in political events.

Calling upon disciplinary methodologies as broad as phenomenology, film studies, and law, each of the chapters is thematically connected by a rigorous attention on the multifaceted ways atmosphere play an important role in the development of shared emotion. While the concept of atmosphere has become a critical notion across several disciplines, the relationship between atmospheres and shared emotion remains neglected. The idea of sharing emotion over a particular event is rife within contemporary society. From Brexit to Trump to Covid-19, emotions are not only experienced individually, they are also grasped together. Proceeding from the view that atmospheres can play an explanatory role in accounting for shared emotion, the book promises to make an enduring contribution to both the understanding of atmospheres and to issues in the philosophy of emotion more broadly.

Offering both a nuanced analysis of key terms in contemporary debates as well as a series of original studies, the book will be a vital resource for scholars in contemporary philosophy, aesthetics, human geography, and political science.

**Dylan Trigg** is an FWF Senior Researcher at the University of Vienna, Department of Philosophy running a project on the phenomenology of nostalgia.

## Ambiances, Atmospheres and Sensory Experiences of Spaces
### Series Editors:
**Rainer Kazig**, *CNRS Research Laboratory Ambiances – Architectures – Urbanités, Grenoble, France*
**Damien Masson**, *Université de Cergy-Pontoise, France*
**Paul Simpson**, *Plymouth University, UK*

Research on ambiances and atmospheres has grown significantly in recent years in a range of disciplines, including Francophone architecture and urban studies, German research related to philosophy and aesthetics, and a growing range of Anglophone research on affective atmospheres within human geography and sociology.

This series offers a forum for research that engages with questions around ambiances and atmospheres in exploring their significances in understanding social life. Each book in the series advances some combination of theoretical understandings, practical knowledges and methodological approaches. More specifically, a range of key questions which contributions to the series seek to address includes:

- In what ways do ambiances and atmospheres play a part in the unfolding of social life in a variety of settings?
- What kinds of ethical, aesthetic, and political possibilities might be opened up and cultivated through a focus on atmospheres/ambiances?
- How do actors such as planners, architects, managers, commercial interests and public authorities actively engage with ambiances and atmospheres or seek to shape them? How might these ambiances and atmospheres be reshaped towards critical ends?
- What original forms of representations can be found today to (re)present the sensory, the atmospheric, the experiential? What sort of writing, modes of expression, or vocabulary is required? What research methodologies and practices might we employ in engaging with ambiances and atmospheres?

**Atmospheres and Shared Emotions**
*Edited by Dylan Trigg*

For more information about this series, please visit: www.routledge.com/Ambiances-Atmospheres-and-Sensory-Experiences-of-Spaces/book-series/AMB

# Atmospheres and Shared Emotions

Edited by
Dylan Trigg

Routledge
Taylor & Francis Group

LONDON AND NEW YORK

First published 2022
by Routledge
2 Park Square, Milton Park, Abingdon, Oxon OX14 4RN

and by Routledge
605 Third Avenue, New York, NY 10158

*Routledge is an imprint of the Taylor & Francis Group, an informa business*

*British Library Cataloguing-in-Publication Data*
A catalogue record for this book is available from the British Library

*Library of Congress Cataloging-in-Publication Data*
Names: Trigg, Dylan, editor.
Title: Atmospheres and shared emotions / edited by Dylan Trigg.
Description: Abingdon, Oxon; New York, NY: Routledge, 2022. |
    Series: Ambiances, atmospheres and sensory experiences of spaces |
    Includes bibliographical references and index.
Identifiers: LCCN 2021026674 (print) | LCCN 2021026675 (ebook) |
    ISBN 9780367674199 (hbk) | ISBN 9780367674205 (pbk) |
    ISBN 9781003131298 (ebk)
Subjects: LCSH: Emotions. | Mood (Psychology) | Atmosphere.
Classification: LCC BF511 .A84 2022 (print) | LCC BF511 (ebook) |
    DDC 152.4—dc23
LC record available at https://lccn.loc.gov/2021026674
LC ebook record available at https://lccn.loc.gov/2021026675

ISBN: 978-0-367-67419-9 (hbk)
ISBN: 978-0-367-67420-5 (pbk)
ISBN: 978-1-003-13129-8 (ebk)

DOI: 10.4324/9781003131298

Typeset in Times NR MT Pro
by KnowledgeWorks Global Ltd.

# Contents

# Figures

# Contributors

**Valeria Bizzari** currently works at the Husserl Archives of the KU in Leuven. Her research interests involve phenomenology, philosophy of emotions and phenomenological psychopathology. Thanks to DAAD and Thyssen fellowships, from February 2018 to September 2020, she worked at the Clinic for General Psychiatry, Universität Heidelberg, with a project on "Asperger's syndrome: A Philosophical and Empirical Investigation of Intersubjectivity and its Disruptions," under the supervision of Professor Thomas Fuchs. She has been a visiting researcher at the Center for Subjectivity Research in Copenhagen, the Oxford Empathy Programme at University of Oxford, and the Department of Philosophy at the University of Wien.

**Gernot Böhme** studied Physics, Mathematics and Philosophy, PhD 1965 Hamburg, Habilitation 1972 Munich. 1969–77 research fellow at the Max-Planck-Institute, Starnberg (Habermas, v. Weizsäcker), 1977–2002 Prof. of Philosophy at the Technical University of Darmstadt, 1997–2002 speaker of Graduate School Technification and Society. Since 2005, he has been Director of the Institut für Praxis der Philosophie and until 2020 he was the President of the Darmstadt Goethe-Association. He is widely regarded as a formative figure in research on atmospheres, and also well-known for his work in Classical Philosophy (Kant, Plato), Social Studies of Science, Philosophical Anthropology, Philosophy of Nature, Aesthetics, Ethics, Goethe, and Theory of Time. His publications in English include *Atmospheric Architectures: The Aesthetics of Felt Spaces* (London: Bloomsbury 2017) and *Invasive Technification: Critical Essays in the Philosophy of Technology* (London: Bloomsbury 2012).

**Tonino Griffero** is full professor of Aesthetics (University of Rome "Tor Vergata"). His most recent books include *Atmospheres. Aesthetics of Emotional Spaces* (Routledge, London-New York 2014); *Quasi-Things. The Paradigm of Atmospheres* (SUNY, Albany, N.Y. 2017); with G. Moretti (ed.), *Atmosphere/Atmospheres. Testing a new paradigm* (Mimesis International, Milan 2018); *Places, Affordances, Atmospheres. A Pathic Aesthetics* (Routledge, London-New York 2019); with G. Francesetti (ed.), *Psychopathology and Atmospheres. Neither Inside nor Outside* (Cambridge Scholar, Newcastle upon Tyne 2019); with M. Tedeschini (ed.), *Atmospheres and Aesthetics. A Plural Perspective* (Palgrave Macmillan, Basingstoke-New York 2019); *The Atmospheric "We". Moods and collective Feelings* (Mimesis International, Milan 2021).

**Julian Hanich** is Associate Professor of Film Studies at the University of Groningen, where he also served as Head of the Department of Arts, Culture and Media from 2017 to 2020. He is the author of two monographs: *The Audience Effect: On the Collective Cinema Experience* (Edinburgh University Press, 2018) and *Cinematic Emotion in Horror Films and Thrillers: The Aesthetic Paradox of Pleasurable Fear* (Routledge, 2010). He co-edited *The Structures of the Film Experience by Jean-Pierre Meunier: Historical Assessments and Phenomenological Expansions* (with Daniel Fairfax, Amsterdam University Press, 2019) and an issue of the journal NECSUS on Emotions (with Jens Eder and Jane Stadler, 2019). With Christian Ferencz-Flatz he was also responsible for an issue of Studia *Phaenomenologica* on 'Film and Phenomenology' (2016). In his research he focuses on film-phenomenology, the collective cinema experience, film and imagination, cinematic emotions, and film aesthetics.

**Veronica Iubei** is a PhD Researcher at the Ruprecht Karl University of Heidelberg. She is at the end of her doctorate, working on an interdisciplinary project on embodiment, atmospheres and psycho-pathology. She is a member of the Section of Phenomenology in the Department of Psychiatry and Psychotherapy, headed by Prof. Dr. Thomas Fuchs. She has a humanistic background, with two degrees in the humanities and a research master's degree in philosophy of mind.

**Joel Krueger** is a Senior Lecturer in Philosophy at the University of Exeter. He works primarily in phenomenology, philosophy of mind, and philosophy of cognitive science: specifically, issues in 4E (embodied, embedded, enacted, extended) cognition, including emotions, social cognition, and psychopathology. Sometimes he also writes about comparative philosophy and philosophy of music.

**Lucy Osler** is a postdoctoral researcher in philosophy on the FWF-funded project Antagonistic Political Emotions at the Center for Subjectivity Research, University of Copenhagen. She recently finished her PhD at the University of Exeter, submitting a thesis on "Interpersonal atmospheres: an empathetic approach." Her current research interests circle around phenomenology, embodiment, empathy, online sociality, critical phenomenology, and phenomenological psychopathology. Recent publications include "Feeling togetherness online: a phenomenological sketch of online communal experiences" (*Phenomenology and the Cognitive Sciences*, 2020), "Taking empathy online" (*Inquiry*, 2021), "Engineering affect: emotion regulation, the internet, and the techno-social niche" (co-authored with Joel Krueger, *Philosophical Topics*, 2019), and "Controlling the noise: a phenomenological account of Anorexia Nervosa and the threatening body" (*Philosophy, Psychiatry, & Psychology*, 2021).

**Andreas Philippopoulos-Mihalopoulos** is an academic/artist/fiction author. He is Professor of Law & Theory at the University of Westminster, and Director of The Westminster Law & Theory Lab. His academic books include the monographs *Absent Environments* (2007), *Niklas Luhmann: Law, Justice, Society* (2009), and *Spatial Justice: Body Lawscape Atmosphere* (2014). His fiction *The Book of Water* is published in Greek and English. His art practice includes performance, photography and text, as well as sculpture and painting. His work has been presented at Palais de Tokyo, the 58th Venice Art Biennale 2019, the 16th Venice Architecture Biennale 2016, the Tate Modern, Inhotim Instituto de Arte Contemporânea Brazil, Arebyte Gallery, and Danielle Arnaud Gallery.

**Thomas Szanto** is Associate Professor in Philosophy at the Center for Subjectivity Research at the University of Copenhagen and PI of the Austrian Science Fund (FWF) project "Antagonistic Political Emotions." His publications include the monograph *Bewusstsein, Intentionalität und Mentale Repräsentation: Husserl und die analytische Philosophie des Geistes* (de Gruyter, 2012), the co-edited volumes *The Phenomenology of Sociality: Discovering the 'We'* (Routledge, 2016) and *The Routledge Handbook of Philosophy of Emotions* (Routledge, 2020). His articles at the intersection of phenomenology, social cognition, social ontology and the philosophy of emotions appeared among others in the journals *Frontiers in Psychology*, *Human Studies*, *Midwest Studies in Philosophy*, *Phenomenology and the Cognitive Sciences*, *Philosophy Compass*, *Synthese*, and *Topoi*.

**Gerhard Thonhauser** teaches philosophy at TU Darmstadt. He was an Erwin Schrödinger Fellow of the Austrian Science Fund associated with the Collaborative Research Centre "Affective Societies" at Freie Universität Berlin. He holds a PhD in philosophy and M.A.s in philosophy and political science from the University of Vienna. His research focuses on social and political philosophy, classic and contemporary phenomenology, and theories of emotion and affectivity. He wrote and edited several books on Heidegger. His latest publications include "A Multifaceted Approach to Emotional Sharing" (*Journal of Consciousness Studies*, 27 (9–10), 2020, 202–227), "Emotional Sharing in Football Audiences" (written together with Michael Wetzels for *The Journal of the Philosophy of Sport*, 46 (2), 2019, 224–243), and "Shared Emotions: A Steinian Proposal" (*Phenomenology and the Cognitive Sciences*, 17 (5), 2018, 997–1015).

**Dylan Trigg** is an FWF Senior Researcher at the University of Vienna, Department of Philosophy running a project on the phenomenology of nostalgia. He earned his PhD at the University of Sussex (2009), MA at the University of Sussex (2005), and BA at the University of London, Birkbeck College (2004). His research focuses on phenomenology (especially Merleau-Ponty and Bachelard), aesthetics, theories of space and place, and embodiment. His articles have appeared in among other places *Human Studies; Emotion, Space and Society; Phenomenology and Cognitive Science;* and *Continental Philosophy Review.* Trigg is the author of several books including *Topophobia: A Phenomenology of Anxiety* (2017); *The Thing: A Phenomenology of Horror* (2014); and *The Memory of Place: A Phenomenology of the Uncanny* (2012). His work has been translated into multiple languages including French, German, Russian, Japanese, Chinese, and Latvian.

**Íngrid Vendrell Ferran** is Heisenberg-Fellow at the Goethe University Frankfurt. Her research interests are philosophy of mind, epistemology, phenomenology and aesthetics. Some of her publications include: *Die Emotionen. Gefühle in der realistischen Phänomenologie* (Akademie 2008); *Wahrheit, Wissen und Erkenntnis in der Literatur* (ed. with Christoph Demmerling, De Gruyter 2014); *Empathie im Film* (ed. with Malte Hagener, Transcript 2017); *Die Vielfalt der Erkenntnis. Eine Analyse des kognitiven Werts der Literatur* (mentis 2018); and *Beauty* (ed. with Wolfgang Huemer, Philosophia 2019).

# Acknowledgements

The contents of this book derive from a conference organised by myself with the cooperation of Hans Bernhard Schmid, which was hosted at the University of Vienna, Department of Philosophy April 25–26, 2019. Allow me to thank all the participants of this event and contributors to this volume, as well as Hans Bernhard Schmid for his support. Additional thanks here to Katherina Krobath and Michaela Bartsch for their vital administrative support. For additional comments on portions of this manuscript, my thanks to Leyla Sophie Gleissner, Audrey Petit-Trigg, and to the members of University of Vienna's, Werkstatt Phänomenologie. The research carried out for this volume was funded by the Austrian Science Fund (FWF) (M2300 Meitner program) and is herein gratefully acknowledged.

*Dylan Trigg, Vienna,*
*August 2021*

# Introduction

## Atmospheres of shared emotion

*Dylan Trigg*

### A sense of togetherness

Following the death of Prince Philip, Duke of Edinburgh on April 9, 2021, the former Prime Minister of the United Kingdom, John Major, suggested that the grief over Prince Philip's death would be an opportune moment to heal frictions within the family. Talking to the BBC, Major remarked that "[t]hey shared emotion. They share grief at the present time because of the death of their grandfather. I think [this] is an ideal opportunity. I hope very much that it is possible to mend any rifts that may exist" (Murray 2021). Irrespective of the veracity of his statement, Major's passing remarks raise a series of philosophically rich questions: What does it mean to share emotion? Must the sharing of emotion involve a shared intentional object as well as a shared affective state? Is shared emotion a short-term event? And, finally, can shared emotion generate the grounds for fostering emotional well-being?

The idea of sharing emotion over a particular event—especially a critical one—is rife within contemporary society. From the fear and apprehension associated with Brexit, to traumas such as the Christchurch shooting, and then to contemporary anxieties tied up with Covid-19, we seem to be living in an age where emotion is manifest not only as an individual affect but as something that saturates our environment, both online and off-line. In one way or another—and to varying degrees—many of us are gripped by a series of emotions, which derive as much from our private experience as they do from the world more broadly. The result of this is that drawing a distinction between individual and shared emotion has become increasingly ambiguous.

In part, this ambiguity is framed by the ubiquity of media in our lives, which is laden with affective value and normative claims about how we ought to be living. The idea of a society or a culture being gripped by a shared emotion is, of course, nothing new. Historically, key events within a given society—revolution, terrorism, sporting victories—tend to generate the foundation for a sense of emotional commonality between people, albeit often on a short-term basis. Yet what this commonality actually consists of

DOI: 10.4324/9781003131298-1

and whether it legitimately qualifies as a case of "shared emotion" has been difficult to clarify, notwithstanding an abundance of research on the topic (cf. Szanto and Moran 2015; Zahavi 2010, 2015; Schmid 2009).

Nowhere are these dynamics and problematics clearer than in the current Covid-19 crisis. One of the unique and salient aspects of the current pandemic is that it has enforced a collective mode of behaviour upon a large portion of society. Seemingly overnight, we have had to collectively regulate how we lead our lives both during and after lockdown. The result is that the majority of people are bound together through a similar set of cultural practices, such as social distancing, staying at home, and wearing masks in public. These actions did not, as it were, evolve slowly over time through cultural practices and traditions; rather, they were enforced from a top-down perspective, meaning a unique set of circumstances has been created that we ourselves as individuals played little part in creating.

Consider here the act of collectively applauding the work of frontline practitioners, a practice that was especially common during the first lockdown in the UK, where it was conceived. Shortly after the first lockdown, a concerted effort was made in several countries to support the work of carers and medical workers on the frontline by applauding their work. In suburban streets and dense urban areas across numerous major cities, people stood at the door to their home, on the balcony of their apartment, or simply at the nearest window, and clapped with enthusiasm together. As a viral event, the initiative was successful, and, indeed, at least in the UK, the act of clapping together was seen by the media as invoking a collective spirit more often associated with wartime.[1]

What is striking, however, is that these expressions of collectivity are not only gestures oriented towards an intentional object (in this case, frontline staff); they are also self-reflexive insofar as they generate a sense of togetherness between people who would otherwise have nothing or little in common. On the surface, the images and videos broadcast in the media during this time depicted a sense of solidarity and togetherness despite (or because) of any differences between social groups. This sense of togetherness was also reproduced at both governmental and state level, thus the British chancellor Rishi Sunak commented: "We want to look back on this time and remember how, in the face of a generation-defining moment, we undertook a collective national effort, and we stood together" (Partington 2020). In a similar sentiment, the Queen also spoke of her hope that "in the years to come everyone will be able to take pride in how they responded to this challenge" (Davies and Savage 2020).

But whether or not collectively applauding the work of frontline practitioners or hoping for a future of collective pride is an actual case of shared emotion or something more like emotional contagion driven by media and political narratives, and undergone by discrete individuals, is not obvious. What is clear, however, is that many of these practices carried out are saturated with a set of recurring affective tonalities. These tonalities are wide-reaching and pervasive. From the collective grief concerning the loss of old ways of life (Weir 2020) to the shared trauma experienced by front-line workers dealing with mass death (Tosone 2021), and then to everyday instances of anxiety, nostalgia, and depression, what is clear that our understanding of Covid-19 has been marked as much by key moments in its development as it has modulations in shared emotion. Yet the abiding emotion of the current era is perhaps not the contagious panic that marked neither the first lockdown nor the collective applause, but instead a more lingering sense of unease, which is harder to define though no less present.

## The concept of atmosphere

The point of departure for the present volume is that the concept of *atmosphere* can play a critical role in accounting for the structure and experience of shared emotion. Over the last 20 years or so, the term "atmosphere" has come under increased philosophical attention (cf. Anderson 2009; Healy 2014; Bille 2015; Bille, Bjerregaard, and Sorensen 2015; Thibaud 2015; and Buser 2017). From a philosophical perspective, the concept has a rich and varied history, most obviously associated with the work of Hermann Schmitz (Schmitz 2011), Gernot Böhme (Böhme 1993, 2017), and more recently by Tonino Griffero (Griffero 2014, 2017). The concept also has an extended influence in other disciplines, especially in human geography and cultural geography (Bjerregaard 2015) and, in particular, concerning the relation between affect theory and atmosphere (Anderson 2014); the role meteorological dimensions play in the affective concept of atmosphere (McCormack 2018); and the structure and experience of atmospheres (Pink and Sumartojo 2018).

While the concept of atmosphere is a diffused one, there are nevertheless a series of characteristics, which are peculiar to much of the research on atmosphere; namely, atmospheres are affective phenomena, which are grasped pre-reflectively, manifest spatially, felt corporeally, and conceived as semi-autonomous and indeterminate entities. This focus on atmospheres as ontologically indeterminate entities is reflected in the language of atmospheres as "quasi-things" (Griffero 2017); "half-entities" (Schmitz 2011); and "quasi-objects" (Böhme 1993).

For all its pervasiveness, atmosphere remains an ambiguous notion, at least in philosophical research. In part, this ambiguity relates to fundamental questions about how we research the concept of atmosphere. How, after all, to approach an atmosphere conceptually? Do we begin by considering the materiality of an atmosphere? Or is it better to begin by considering how people and groups engender the production of an atmosphere over time? In the same measure, just as the concept of atmosphere raises fundamental questions concerning method, perhaps the ambiguity is not so much a fault but an invariant feature of atmospheres themselves. Indeed, much of the literature on atmosphere affirms the concept as a fuzzy and liminal one.

Consider here the range of things we describe as atmospheric: from a person to a place, from a political situation to an entire epoch, the term "atmosphere" covers an immense terrain. We also speak about works of art as having a particular atmosphere as well as specific events, not least current situations like a pandemic. In each case, it is often difficult to pinpoint where an atmosphere begins and ends, both spatially and temporally. In this respect, an atmosphere often has an "excessive" quality to it, such that it seems to seep into things of its own volition.

This sense of an atmosphere as having an excessive quality to it is clear enough in how atmospheres can often act upon us. Consider here the experience of walking into a room and suddenly and forcefully being overwhelmed by a specific atmosphere, whether it be calming or tense. Consider also the sense of a city as having a peculiar atmosphere of its own, such that we often have to acclimatise ourselves to its singular feel. The atmosphere is not "activated" upon our arrival, but instead belongs to the place and inheres in the materiality of the environment. Thus, when we leave that city, place, or room, the atmosphere does not leave with us. Rather, it remains embedded in the things of the world, such that other people can attune themselves to that atmosphere long after we have left.

Consider, finally, how an atmosphere can be fabricated through design (as in the creation of an inviting atmosphere through ornaments, architectural design, and lighting). Understood in this respect, the material dimension of an atmosphere institutes itself within society with an efficacious power (as evident politically in the manipulation of an atmosphere of fear to achieve a given end). For this reason, an atmosphere can be engineered with the aim of eliciting a certain affective response from a group of people (as in the case of the atmosphere of a rally to reinforce a group's political commitment to a cause). In each case, in acting upon us, atmospheres can often urge us to act in a particular way, reinforcing the sense that far from existing solely in one's mind, emotions are in fact "poured out spatially" (Schmitz 2011, 257).

Because of these multiple ambiguities and indeterminate traits (i.e., "quasi-things" (Griffero 2017); "half-entities" (Schmitz 2011); and "quasi-objects" (Böhme 1993)), atmospheres need multiple (re)conceptualisations, which this volume proposes to offer. In particular, this volume proposes to situate the concept of atmosphere in a wide-reaching methodological framework, which is nevertheless rooted in phenomenology. Yet instead of approaching this task from an abstract or general perspective, the chapters collected here will consider the salient attributes of atmosphere through the concept of shared emotion. From conceptual analysis (Griffero; Thonhauser; Vendrell Ferran) to the expression of shared emotion in atypical phenomena (Trigg; Bizzari and Iubei; Krueger), and then to the aesthetic and political aspects of atmosphere (Hanich; Böhme; Osler and Szanto), this volume covers a broad range of issues, each of which is nevertheless unified in contending with the strengths and limits of atmosphere as a conceptual entity.

## What is shared emotion?

Surprisingly, amongst the key recent works on the phenomenology of atmosphere, including texts by Schmitz (2011), Anderson (2009), Böhme (2017), and Griffero (2014), there has been very little attention given to the relation between shared emotion and atmosphere, with the case of a few exceptions (cf. Slaby 2014). This oversight is especially notable, given that alongside the concept of atmosphere, the adjoining idea of shared emotion has also been a topic of renewed interest in the last decade as well as being topical in and of itself. Indeed, cases of shared emotion are *prima facie* ubiquitous in contemporary life insofar as we freely speak of collective fear (Markus and Kitayama 1994), or shared happiness (Uchida and Oishi 2016). Moreover, everyday speech reinforces the sense that some affective states are more shareable than others, as when people speak of a collective sense of elation regarding a sporting victory or a felt sense of disappointment following the outcome of an election.

In analytical philosophy, the topic of shared emotion often appears within discussions of collective intentionality (Nussbaum 2003; Pacherie 2002). Collective intentionality is the idea that two or more people's minds can be jointly directed towards a given phenomenon. One initial context where the discussion of shared emotion and collective intentionality appears is within debates in analytic philosophy of mind. Much of this research is attentive to how intentionality is not only an act undertaken by an individual, but something that can also involve a strong social dimension, such that intentional states can be shared. Debates in this area have tended to focus on how it is possible for individuals to do something together, especially everyday tasks such as walking together (Gilbert 2013; Velleman 1997).

The roots of these debates can be sourced in Wilfred Sellars while the term "collective intentionality" itself stems from John Searle's influential paper, "Collective Intentions and Actions" (Searle 1990). Central thinkers within these debates have each offered their own take on collective intentions, but what remains consistent through this research is a focus on whether or not collective acts are ultimately reducible to an aggregation of individual actions (Chant, Hindriks, Preyer 2014). Thus, collective intentionality, as it is conceived in analytical philosophy, tends to privilege questions concerning the ontology of collective acts, the psychological states of individuals engaged in those acts, and how each of these dimensions contributes to social institutions (Tuomela and Miller 1988). Critically, what is omitted from this research is the affective, embodied, and cultural structure of intentionality (Pacherie 2002). These elements are important because they (i) demonstrate how cognitive processes are shaped by affective states; (ii) underscore how the body plays a critical role in interpersonal communication; and (iii) reveal how emotions are mediated by sedimented historic and cultural sensibilities embedded in the world.

Today, research on shared emotion is blossoming, such that there is a "motley of overlapping phenomena that do not make up a single natural kind [of shared emotion]" (Michael 2016). Despite this fertile activity, the phenomenological heritage dating back to Husserl of dealing with shared emotion has largely been overlooked (cf. Szanto and Moran 2015). Indeed, already in Scheler's *The Nature of Sympathy*, four forms of "fellow feeling" are delineated—"feeling together," empathy, sympathy, and "affective contagion"—which only now are beginning to be recognised in anticipating recent debates in philosophy of mind (Scheler 1913/1954). The current volume proceeds from the conviction that phenomenology can offer a convincing—though, not absolute—alternative to analytic treatments of intentionality and emotion in at least two respects.

In the first case, phenomenology is uniquely suited to attending to how shared emotion emerges through a first-person perspective. Such a perspective is vital for clarifying how individual existence is social from the outset. This idea has been almost largely overlooked in other traditions, and it is only recently that analytical philosophy has considered the importance of "social ontology"—a term that, in fact, derives from phenomenology (Husserl 1970, 2012). Indeed, the idea of collective intentionality itself is traceable to early phenomenologists such as Gerda Walther (1923), Edith Stein (1917), and Max Scheler (1954). This legacy is of interest, not only for historic reasons, but also in terms of shedding light on the experience and structure of shared emotion.

In the second case, phenomenology issues a challenge to analytic treatments of intentionality and collective intentionality in their narrow focus on joint action, team reasoning, and shared agency. Phenomenology brings to these debates a much wider focus on affectivity, embodiment, as well as cultural, political, and social dimensions, which is indispensable for an investigation of shared emotion. These dimensions are neither contingent nor incidental in the analysis of shared emotion; rather, phenomenology demonstrates their significance not only for shared emotion itself but also for advancing understanding in key socio-political events, as this volume testifies.

## The context for the book

The contributions in this book derive from a two-day international conference, held at the University of Vienna, Department of Philosophy from April 25 to 26, 2019 and organised by myself with the co-operation of Hans Bernhard Schmid, which explored the intersection between atmospheres and shared emotions. The chapters collected here are not simply reprints of the conference findings; rather, they are elaborations, developments, and, in some cases, entirely new forays into a dynamically expanding field of research.

Moreover, although the title of this book brings together atmospheres and shared emotion, the issue of shared emotion varies throughout. While some chapters treat shared emotions on an explicit level (see especially Griffero; Bizzari and Iubei; Hanich; Osler and Szanto), other chapters interrogate the foundations of what renders shared emotion possible (see especially Thonhauser; Vendrell Ferran; Trigg; Philippopoulos-Mihalopoulos). This diversity in approach is mirrored by the range of topics covered. Alongside conceptual analysis (Griffero; Thonhauser; Vendrell Ferran), several chapters focus on specific articulations of shared emotion in a wide range of phenomena, ranging from psychopathology (Bizzari and Iubei) to cinematic audiences (Hanich), and to autism (Krueger).

Throughout, phenomenology is the methodological thread around which these chapters revolve. In the same measure, however, as editor of this volume, I am gratified that the phenomenology in this book is a plurality of *phenomenologies* rather than a monolithic (not to say tedious) formulation of phenomenology as a singular kind of (classical) methodology. This plurality of methods is clear enough in the scope of the chapters. While some contributors robustly defend phenomenology as an integral mode of researching atmospheres (Griffero; Bizzari and Iubei; Hanich), other chapters evince a critical relationship to the method (Philippopoulos-Mihalopoulos; Krueger) in terms of its potential (and sometimes propensity) for excluding and neglecting under-represented perspectives. In addition, alongside methodological matters, the thematic focus on atmosphere is also given a wide range of expressions. As the volume makes clear, atmosphere can range from the foundation of shared emotion (Griffero), which can generate insight into affective disorders (Trigg), to a site of exclusion (Krueger) marked by a series of conceptual and ideological blind spots (Philippopoulos-Mihalopoulos). As such, atmospheres are not only forces of benevolence (Hanich), they can also be used for pernicious (Böhme) if not manipulative (Osler and Szanto) ends.

Let me finally add that as editor, I am especially gratified to bring together many of the leading scholars in the field of atmosphere research, as well as an emerging generation of thinkers who are already redefining how we think about atmospheres both conceptually and otherwise. As researchers on atmosphere—as veritable *atmospherologists*—this volume attests to the complexity of atmosphere in both its historical and conceptual understanding. With this said, I now present an overview of the chapters.

## Outline of the book

### Moods and atmospheres

*Atmospheres of Shared Emotion* begins by investigating the salient attributes of an atmosphere from a conceptual perspective. Central to this work is both the distinction between atmospheres and moods, as well as the question of how an atmosphere generates the foundation for shared emotion to be possible in the first place. Accordingly, the first chapter by Tonino Griffero investigates what role atmospheres play in the production of shared emotion. Griffero begins this study by asking what it means to live in a shared space. As he argues, we take this idea as being obvious, yet this taken-for-granted quality tends to conceal the complexity embedded in the idea of shared emotion. For example, while two people can be attuned to an atmosphere and recognise the atmosphere as actually existing, they can nevertheless interpret the same phenomena in wildly different ways. In a careful analysis, he explores this ambiguity, locating a series of doubts raised against the idea of shared emotion before positing what he terms an "atmospherologic perspective." At the heart of this approach is a fine-grained analysis of the relational structure of a shared atmosphere, with a specific focus on how felt-bodily processes shape and determine the appearance of an atmosphere.

The following chapter by Gerhard Thonhauser compliments Griffero's chapter, insofar as Thonhauser's concern is with the history of the term Stimmung. Along with Griffero's chapter, this chapter is critical for explicating and analysing key concepts within debates on atmosphere. Central to Thonhauser's study is the question of how mood relates to atmosphere. His chapter is at once a historical and conceptual analysis of the development of the term Stimmung, yet in the same measure, this analysis also impacts the understanding of Stimmung as an experience. This is clearly evident in Thonhauser's placing of Stimmung within the context of aesthetics, especially artworks and landscapes. Indeed, building on these textual foundations, his chapter explicates how the concept of Stimmungen serves as a communal function, facilitating the communality of various modes of affectedness, and thereby establishing corresponding social collectives.

Offering a counterpart to Thonhauser's chapter, the chapter by Íngrid Vendrell Ferran offers a subtle yet probing analysis of the distinction between atmospheres and moods. She begins this investigation with the observation that in everyday speech, we tend to employ the same terms to refer to moods and to atmospheres: we claim to be cheerful, depressed, and melancholic, and we claim that an object or a situation is presented as cheerful, depressed, and melancholic. This raises the question: are moods and atmospheres interchangeable, and if not, then why do we employ the same terms? Vendrell Ferran responds to this question through undertaking a careful analysis of the semantic and conceptual structure of these terms, before defending what she calls the "distinctiveness thesis." As she sees it, the key distinction between feeling an emotion and perceiving an emotion within a given surrounding is experiential. As such, through unpacking the salient distinctions between these modes of experience, Vendrell Ferran generates a clarified but also renewed understanding of what an atmosphere itself consists of.

## Psychopathological atmospheres

The next chapter marks a shift from conceptual analysis of key terms to an examination of how the concept of atmosphere can shed light on psychopathological conditions and affective disorders. Accordingly, Dylan Trigg investigates the anxiety involved in the current pandemic crisis. The point of departure for this chapter is the idea of conceiving anxiety in atmospheric terms. This stands in contrast to how the emotion is conventionally understood, which is either as a cognitive state or as embodied but nevertheless subjective phenomena. Approaching anxiety as an atmosphere means recognising how the structure of the world (and not just our mental or bodily states) is shaped by affective moods. This is especially clear in the case of Covid-19. By employing atmosphere as a key concept, it is possible to show how a phenomenon such as Covid-19 is not just a discrete disease to be understood in medical terms; more complex than this, it is also an affective atmosphere that forges new modes of comportment in the world. Two specific ways this presents itself is through a revised understanding of the home and of the lived body, with both revealing themselves as sites of familiarity and anxiety.

In a co-authored chapter, Valeria Bizzari and Veronica Iubei continue the theme of atmospheric disturbances by accounting for the existence of what they term "atmospheric intercorporeality." In their view, atmosphere is the vehicle that shapes the affective life and the subjective communication with the lifeworld. Together, Bizzari and Iubei unpack this claim in two stages. First, they critically introduce the notion of atmosphere, emphasising its role in the development of sociality (from intercorporeality to shared emotions). Second, they draw upon the case study of a schizophrenic patient to explicate the idea of atmosphere as an intercorporeal relation. In doing so, they reveal how schizophrenic subjects not only lose empathic attunement with others but also undergo a disruption at a deeper level of corporeity. Their chapter thus sheds light on the role atmosphere can play not only in accounting for felicitous instances of life, but also in terms of illuminating the structure of psychiatric disorders.

Ending the section on psychopathological atmospheres, Joel Krueger's chapter marks a critical axis in the volume in several respects. Krueger's aim in his chapter is to turn from ontological questions concerning *what* atmospheres are to the equally pressing issue of what do atmospheres *do*? In particular, he focuses on how atmospheres can shape experience in both an inclusionary and exclusionary way. To give expression to this thesis, he turns to both the critical phenomenology Sara Ahmed as well as to social difficulties in autism. What these trajectories of thought have in common is that they disclose the normative foundations from which atmospheres evolve. For example, in his analysis of the work of Ahmed, Krueger shows how atmospheres, as affective arrangements, can generate spaces that allow bodies to fit and extend into them in a variety of ways. In our everyday life, atmospheres are typically presented as spaces of belonging, in which individuals and groups fit in *together*. Yet this taken-for-granted certitude conceals the ways in which atmospheres can also be disorientating and disquieting. By looking at Ahmed's notion of "being stopped," emblematically taken up in the form of non-white bodies being stopped by police, Kruger carefully unpacks the presuppositions sedimented in our habitual understanding of atmosphere, which are underpinned at all times by political, racial, and gendered dimensions (see also Osler and Szanto's chapter). Building on this foundation, Kruger then applies this framework to the case of autism. Here, too, an atmosphere of exclusion is forged, which tends to frame autistic modes of embodiment as "unusual or strange." Kruger's chapter thus reveals that a critical phenomenological approach to atmospheres is not only integral in and of itself, it is also beneficial in shedding light on salient aspects of the notion of atmosphere more broadly.

### Aesthetic and political atmospheres

The final part of the book combines political atmospheres with aesthetic atmospheres. Far from divergent topics, as the chapters in this section reveal, aesthetic manipulation is central to the structure of political atmospheres. Beginning this section, Julian Hanich examines the role shared atmospheres play in cinema audiences. In particular, he generates an analysis of the different modalities of collective laughter at work in the cinema. In undertaking this analysis, he argues that the staging of atmospheres does not only derive from a creator; rather, it is co-constituted between the designer (in this case, the filmmaker) and the audience. In conjunction, Hanich simultaneously posit what he calls *spread collective emotions*. With this, he makes a subtle distinction between how we experience shared emotion together and how an emotion can pour itself over a group without that emotional spread resulting in any kind of collective experience. This insight is critical because it contributes to clarifying to what is specific and singular to shared emotion that is absent in other but related affective phenomena.

From comedy to tragedy, Gernot Böhme's chapter brings together the theme of aesthetics and politics by looking at how National-Socialist's design for mass events and architecture stimulates collective emotions, in turn, producing a sense of loyalty to the regime. Böhme's examples are taken from mass assemblies like the Parteitag in Nuremberg, from the so-called Ordensburgen, and the architectural design of a New Berlin. Here, Böhme focuses especially on reports provided by Denis de Rougemont of the emotional effects of Hitler's performances as well as the architecture of Albert Speer. The theoretical background for analysing architecture as communicative design is given by the results of New Phenomenology, in particular, by Hermann Schmitz's concept of Eindruckstechnik, the technology of producing emotional impressions to people. As a conclusion, Böhme consolidates his findings by suggesting that engendering atmospheres may be used to manipulate people as to their hopes, fears, and desires.

The ninth chapter, by Lucy Osler and Thomas Szanto, focuses on the structure of political atmospheres and, in particular, what makes an atmosphere political. The issue of political atmospheres is at the foreground of contemporary research on atmospheres, and Osler and Szanto's chapter makes an important contribution in advancing this field in several areas. Their aim is to analyse and understand political atmospheres through developing what they term an "interpersonal atmosphere." With this, they put forward a relational model of atmosphere that can provide a key explanatory role in accounting for the variances at work in political atmospheres. As with Krueger's chapter, Osler and Szanto are as much interested in what atmospheres *are* as what they *do*. More specifically, their chapter leads to an analysis of the role normative factors play in the structure of shared emotion, especially in terms of how these normative dimensions consolidate integration within a group. This normative dimension also sheds light on how emotions are regulated within a group, thereby clarifying how political atmospheres are not only affective states, but are also social modes of comportment.

The final chapter, by Andreas Philippopoulos-Mihalopoulos, serves as a conclusion to the volume. In it, Philippopoulos-Mihalopoulos takes a provocative—even, polemical—look at the concept of atmosphere precisely as it is deployed by "atmospherologists" themselves, including those who contribute to this volume. At the heart of his critical analysis is the question of what role atmosphere plays in the eco-system of atmospherologists, atmosphere studies, and the broad presentation of atmosphere as being a specific kind of entity; namely, ineffable, fuzzy, and framed as a potential space of belonging at the expense of excluding otherness. In undertaking this study, he simultaneously interrogates what function an atmosphere would have if its "ineffable" character was finally "clarified" (if, indeed, such a clarification was possible). In this respect, Philippopoulos-Mihalopoulos brings to light a psychoanalytical and post-human approach to the literature on atmospheres, which is often populated by a specific form of phenomenological humanism. His chapter, thus, not only targets the usage of atmosphere within atmosphere studies, but also raises a series of tacit questions about the phenomenological methodology more broadly and, in particular, the extent to which phenomenological research is determined in advance. While deconstructive in nature, his reflections resist being destructive. Indeed, at the end of his chapter, Philippopoulos-Mihalopoulos presents a constructive challenge to the issue of shared emotion; namely, to collectively withdraw from atmospheres in the hope of establishing a space of reciprocity for otherness, such that emergent and overlooked perspectives finally come into view.

Needless to say, firm resolutions to these questions—and indeed, to the issues raised in the volume more broadly—do not exist in an unequivocal form. In addition, it should be noted that while research on atmosphere is thriving, especially from a phenomenological perspective, there is a notable lacuna with respect to the research on gender-bias in atmospherics. In part, this oversight is not so much a question of the under-representation of gender within research on atmospheres, but instead points to the question of to what extent gender plays a role in the construction and emergence of atmospheres. While individual chapters in this volume gesture towards these themes (esp. Krueger and Philippopoulos-Mihalopoulos), it is clear that more research is needed. The hope of this volume is that in raising questions concerning the fundamental structure of atmosphere, as well as exploring the multifaceted expression of atmospheres within affective life, the conceptual terrain marking the research on atmospheres will be refined and enriched through the addition of these chapters.

## Note

1  It should also be noted here that as successful as this initiative was as a social media event, it also risked being a performance of shared emotion with any content. Indeed, the fact the organiser of the event ultimately expressed a desire to put an end to it supports this view.

## References

Anderson, Ben. 2009. "Affective atmospheres." *Emotion, Space, and Society*, 2: 77–81.

Anderson, Ben. 2014. *Encountering Affect*. London: Routledge.

Bjerregaard, Peter. 2015. "Dissolving objects: museums, atmospheres and the creation of presence." *Emotion, Space, and Society*, 15: 74–81.

Bille, Mikkel. 2015. "Lighting up cosy atmospheres in Denmark." *Emotion, Space and Society*, 1: 56–63.

Bille, Mikkel, Bjerregaard, Peter & Sorensen, Tim. 2015. "Staging atmospheres: materiality, culture and the texture of the in-between." *Emotion, Space and Society*, 15: 31–38.

Böhme, Gernot. 1993. "Atmosphere as the fundamental concept of a new aesthetics." *Thesis Eleven*, 36: 113–26.

Böhme, Gernot. 2017. *Atmospheric Architectures: The Aesthetics of Felt Space*. London: Bloomsbury.

Buser, Michael. 2017. "The time is out of joint: Atmosphere and hauntology at Bodiam Castle." *Emotion, Space and Society*, 25: 5–13.

Chant, Sara, Hindriks, Frank & Preyer, Gerhard. 2014. *From Intentionality to Collective Intentionality*. Oxford: OUP.

Davis, Caroline & Savage, Michael. 2020. "Queen to tell nation to 'take pride' in response to pandemic." *Guardian*, April 4, 2020. https://www.theguardian.com/world/2020/apr/04/queen-to-make-personal-tv-address-to-the-nation-on-sunday

Gilbert, Margaret. 2013. *Joint Commitment: How We Make the Social World*. Oxford: OUP.

Griffero, Tonino. 2014. *Atmospheres: Aesthetics of Emotional Spaces*. Trans.SaraDe Sanctis. New York: Routledge.

Griffero, Tonino. 2017. *Quasi-Things: The Paradigm of Atmospheres*. Trans. Sara De Sanctis. New York: State University of New York Press.

Healy, Stephen. 2014. "Atmospheres of consumption: shopping as involuntary vulnerability." *Emotion, Space and Society*, 10: 35–43.

Husserl, Edmund. 1970. *The Crisis of European Sciences and Transcendental Phenomenology*. Trans. David Carr. Evanston: Northwestern University Press.

———. 2012. *Ideas: General Introduction to Pure Phenomenology*. Trans. W.R. Boyce Gibson. London: Routledge.

Markus, Hazel. & Kitayama, Shinobu. 1994. "A collective fear of the collective: Implications for selves and theories of selves." *Personality and Social Psychology Bulletin*, 20: 568–579.

McCormack, Derek. 2018. *Atmospheric Things: On the Allure of Elemental Envelopment*. Durham: Duke University Press.

Michael, John. 2016. "What are shared emotions (for)?" *Frontiers in Psychology*. https://doi.org/10.3389/fpsyg.2016.00412

Murray, Jessica. 2021. "Prince Philip's death 'ideal opportunity' to heal royal rifts, says John Major." *Guardian*, April 11, 2020. https://www.theguardian.com/uk-news/2021/apr/11/prince-philips-death-ideal-opportunity-to-heal-royal-rifts-says-john-major

Nussbaum, Martha. 2003. *Upheavals of Thought: The Intelligence of Emotion*. Cambridge: Cambridge University Press.

Partington, Richard. 2020. "UK government to pay 80% of wages for those not working in coronavirus crisis." *Guardian*, March 20, 2020. https://www.theguardian.com/uk-news/2020/mar/20/government-pay-wages-jobs-coronavirus-rishi-sunak

Pacherie, Elisabth. 2002. "The role of emotions in the explanation of action." *European Review of Philosophy*, 5: 55–90.

Pink, Sara & Sumartojo, Shanti. 2018. *Atmospheres and the Experiential World: Theory and Methods*. London: Routledge.

Scheler, Max. 1954. *The Nature of Sympathy*. London: Routledge.

Schmid, Hans Bernhard. 2009. *Plural Action: Essays in Philosophy and Social Science*. Dordrecht: Springer.

Schmid, Hans Bernhard. 2009. "Shared Feelings. Towards a Phenomenology of Collective Affective Intentionality." In H.B. Schmid, *Plural Action*. Dordrecht, Springer.

Schmitz, Hermann. 2011. "Emotions outside of the box—the new phenomenology of feeling and corporeality." *Phenomenology and the Cognitive Sciences*, 10: 241–259.

Searle, John. 1990. "Collective Intentions and Actions." In Cohen, Philip R., Morgan, Jerry, & Pollack, Martha E., eds., *Intentions in Communication*, 401–415 Cambridge Mass: Bradford.

Slaby, Jan. 2014. "Emotions and the Extended Mind." In: Salmela, M., & von Scheve, C., eds., *Collective Emotions*. Oxford: Oxford University Press.

Stein, Edith. 1917. *Zum Problem der Einfühlung. Edith Stein Gesamtausgabe, Vol. 5*. Wien, Basel, Köln: Herder 2008. [*On the Problem of Empathy*. Transl. by W. Stein. Washington, D.C.: ICS Publication 1989.]

Szanto, Thomas & Moran, Dermot. 2015. *Phenomenology of Sociality: Discovering the We*. New York: Routledge.

Thibaud, Jean-Paul. 2015. "The backstage of urban ambiances: when atmospheres pervade everyday experience." *Emotion, Space and Society*, 15, 39–46.

Trcka, Nina. 2017. "Collective moods. A contribution to the phenomenology and interpersonality of shared affectivity." *Philosophia*, 45(4): 1647–1662.

Tosone, Carol. 2021. *Shared Trauma, Shared Resilience During a Pandemic: Social Work in the Time of COVID-19*. Dordrecht, Springer.

Tuomela, Raimo & Miller, Kaarlo. 1988. "We-Intentions." *Philosophical Studies*, 53: 115–137.

Uchida, Yukiko & Oishi, Shigehiro. 2016. "The happiness of individuals and the collective." *Japanese Psychological Research* 58.1: 125–141.

Velleman, David. 1997. "How to share an intention." *Philosophy and Phenomenological Research* 57 (1): 29–50.

Walther, Gerda. 1923. "Zur ontologie der sozialen gemeinschaften." *Jahrbuch für Philosophie und phänomenologische Forschung* 6: 1–158.

Weir, Kirsten. 2020. *Grief and COVID-19: Mourning our bygone lives*. April 1, 2020. *American Psychological Association*, https://www.apa.org/news/apa/2020/04/grief-covid-19

Zahavi, Dan. 2010. "Empathy, embodiment and interpersonal understanding: from Lipps to Schutz." *Inquiry*, 53(3), 285–306.

—— 2015. "You, me and we: the sharing of emotional experiences." *Journal of Consciousness Studies* 22 (1–2): 84–101.

# Part I
# Moods and atmospheres

# 1    Are atmospheres shared feelings?

*Tonino Griffero*

## What is "sharing" atmospherically?

Everyone occasionally gets the cogent impression of living in a shared atmosphere. But when you question the meaning of this sharing, this certainty is lost and affective sharing seems to be able to live with, somewhat paradoxically, different feelings. In D.H. Lawrence's *The Rainbow*, for example, the "same" atmosphere of the Lincoln Cathedral is experienced in a very different way by two characters. Both are overcome with wonder and awe and clearly feel the powerful *genius loci*, but whereas for Will "'before' and 'after' were folded together, all was contained in oneness" in a "timeless ecstasy," and the church "was all, this was everything," Anna is instead "silenced rather than tuned to the place," so that Will's "passion in the cathedral at first awed her, then made her angry." More precisely, she feels that, compared to the "pillars upwards [...] there was the sky outside," and this

> open sky was no blue vault, no dark dome hung with many twinkling lamps, but a space where stars were wheeling in freedom, with freedom above them always higher. The cathedral roused her too. But *she would never consent* to the knitting of all the leaping stone in a great roof that closed her in, and beyond which was nothing, nothing, it was the ultimate confine [...] Her soul too was carried forward to the altar, to the threshold of Eternity, in reverence and fear and joy [...] But even in the dazed swoon of the cathedral, *she claimed another right*. The altar was barren, its lights gone out. God burned no more in that bush. It was dead matter lying there. *She claimed the right to freedom above her, higher than the roof. She had always a sense of being roofed in.*
>
> (Lawrence 1995, 147; my emphasis)

This passage, here cited to demonstrate how a certain deeply involving atmosphere can also elicit an opposite emotional reaction, shows, however, that the atmosphere is clearly the "same" (wonder and awe). What changes is that Anna "wanted to get out of this fixed, leaping, forward-travelling

DOI: 10.4324/9781003131298-2

movement," for example, by reaching "at little things, which saved her from being swept forward headlong in the tide of passion that leaps on into the Infinite in a great mass, triumphant and flinging its own course" (147). By means of a growing resistance based on indulging on the details, which act here as discrepant and disturbing sub-atmospheres, she maliciously succeeds in

> spoiling his passionate intercourse with the cathedral and destroying the passion he had [...] Strive as he would, he could not keep the cathedral wonderful to him. *He was disillusioned.* That which had been his absolute, containing all heaven and earth, was become to him as to her, a shapely heap of dead matter-but dead, dead (148).

In this case, for both the first atmospheric impression changes over time: for her, it does so autonomously and thanks both to discrepant sub-atmospheres and to her increasing need to rebel against a too authoritarian and flagrantly "manipulative" atmosphere; for him, because of the powerful negative atmosphere she radiates on him. It follows that, if in the *Lebenswelt* an atmosphere is first of all the authoritative "object" of natural perception, it is also sometimes filtered through the ideas and evaluations of the perceiver and is even an invitation that, under certain circumstances, can be changed or partly declined.

In most cases, in our everyday life, atmospheres exist in a less objective form (so to speak), as they are placed "between" the two necessary poles of object (or rather, the environmental *qualia*) and subject (or rather, their felt-body). This flexibility though should never be overestimated to the point of confusing the diversity of affective and felt-bodily reactions to an atmospheric feeling (of the way different perceivers differently "filter" it) with the diversity of the atmosphere perceived. This instead—precisely because it is relatively the same—allows for the actual coexistence (and even the intersubjective communicability) of relatively different resulting moods. But, I will come back to this later. For now, it is enough to reiterate that often we seem to "breathe" the same atmosphere and that saying that an atmosphere is a shared feeling sounds a bit like a tautology.

Needless to point out how puzzling this thesis is. Especially once the transcendental, non-conceptual universality that Kant attributed to feelings such as that of the beautiful has become unsustainable. In fact, how can one be sure that this over-intellectualised delight, as Kant would assume, is so independent of personal interest (in the broad sense) as to be universally presupposed in equally free and uninterested people? Even a relic like the Kantian transcendental argument, however, can still be instructive. Indeed, it suggests that sharing feelings do not derive from empirical-statistical evidence or their propagation through chain reactions, but on simply acknowledging an affective "we" that is irreducible to empirical-social intersubjectivity.

The so-called "affective turn," powered today by the obvious but theoretically long-neglected consideration that humans do more than think and reason in a logical way, does not consist primarily of inner and private states or processes.[1] Hence, the current tendency to promote the idea that, especially within our very stratified and post-conventional society, humans can feel "with" others exactly as they can share beliefs and intentions:[2] this idea obviously goes against the traditional theoretical but also common-sensical view of affective individualism[3] according to which feelings are necessarily "internal" to an individual.

For my atmospherology (cf. Griffero 2014a, 2017a, 2019a, 2020a, 2021)[4] and, in general, for every atmospheric approach,[5] the question of collective feelings, as already mentioned, seems a self-evident implication and, at the same time, a real "elephant in the room." In fact, it could even be said that if atmospheres[6] are conceived as feelings pervading and tonalising a certain space, then they are something shared by definition. It is quite clear that by saying, for example, that a certain room is oppressive or relaxing, or that (as in an old popular song) "love is in the air, everywhere I look around," we are implicitly acknowledging that a certain feeling, instead of being groundless like *Stimmungen* or just quick-intentional like emotions, is now poured out into a (lived) space.[7] And exactly for this reason is quasi-objectively experienced in its authority by everyone entering (or staying in) that space.[8] Like any "true" perceived, an atmosphere imposes its presence in public as a shared "object" (cf. Wiesing 2014). Nevertheless, this line of thought is too simple and needs further investigation.

The first thing to notice is the very ambiguity of the notion of sharing feeling. Consider phrases like "we know just what someone is feeling," "we feel for someone," "we feel with them," "we empathize," "we imagine how they feel," "we put ourselves in their shoes," "we sympathize," "we resonate to their feelings," "our heart goes out to them," etc. (cf. Goldie 2009). First of all, it is not at all clear whether, for example, a subjective-projective feeling (what I call a spurious atmosphere) can be shared in the same way as a real environmental feeling; whether a distinction should be made between more or less sharable feelings (respectively, shame or pride); whether a feeling could be really shared without a supra-individual subject as well as a structure that, due to its vagueness and longer duration, does not afford the fast way to think and act provided by emotions.[9]

Nevertheless, everyday experience offers many examples of situations in which people gathered on a certain place express collectively and sometimes involuntarily their shared feeling by shouting, chanting, gesturing, etc. What's more: our entire affective life, namely what we feel as well as how we come to feel, is essentially scaffolded by collective atmospheric situations, just as the affective development of new-borns is scaffolded by familiar atmospheres optimised or regulated by their caregivers.[10] This means that someone's endogenous "subjective facts" (to use a neo-phenomenological notion), which, unlike the "objective facts," result from their first-personal and felt-bodily resonances[11] and one can only refer by using one's own name, are even if not mainly exogenously determined by synchronical and diachronical external atmospheric feelings. However, seeing in this atmospheric "we-space" a set of feelings extended beyond the confines of someone's physical body and resulting from continuously updated transactions between the self and others (even if only imagined or forecasted),[12] would not be a ground-breaking stance. Indeed, speaking of a shared intercorporeality (*à la* Merleau-Ponty) without ever specifying the grammar and syntax of a far from obvious felt body alphabet is little more than a metaphorical description of intent.[13]

Atmospheres are not only external tools like the affordances of music and venues, the expressions and behaviours of individuals or groups, rituals, habits, and other people's feelings. They also function both as scaffolds of affective experiences that would not be realisable without actions that help solidify hitherto inchoative emotional experiences, thereby establishing some sort of socio-normative affective appropriateness (cf. Slaby 2014). Nevertheless, they are, at least in the case of what I call "prototypic atmosphere," something ontologically more than "tools for feeling" (Slaby 2014, 43), since they are feelings that are literally spread out in space and in most emblematic cases act as quasi-things[14] felt-bodily involving individuals and groups and resisting any person's effort to change or neutralise them. The atmospheric externalism is, therefore, more radical than the so-called environmentally and socially emotion-externalism.[15] Precisely, because of their relative independence from whether and how they are perceived, atmospheres seem perfectly entitled to reclaim a leading role in the current debate on shared feelings.

We must not forget, however, that atmosphere sharing is only made possible by ubiquitous felt-bodily communication.[16] In fact, as a sounding board for atmospheres, our felt body makes possible thinking of every perception as a felt-bodily communication based on a certain relationship between incorporation and excorporation. It communicates with extra-organismic *Gestalten* (animate or inanimate) and their affordances, more specifically with motor suggestions and synaesthetic characters. This kind of communication, not confined to physical body[17] and, therefore, fully engaged with the external world, explains better than a generic common humanity how and why an atmosphere can be collective and shared. Indeed, the felt-bodily dynamic underlying experience or perception gives life to a peculiar ad-hoc felt body each time, thus providing the fundamental form of world-disclosure that Heidegger assigned to (unfortunately too unbodied) *Stimmungen*.[18] Since, however, individuals feel and realise an atmospheric feeling only in their own lived bodies, the assumption of an ad-hoc collective felt body does not involve the assumption of a collective ontological subject. Atmospherologically speaking, in fact, the collective subject gives way to an atmospheric quasi-thing that is, at least initially, relatively independent from participating subjects.

## Serious doubts

Nevertheless, this context gives rise to many doubts. Here are four examples. The first question concerns the semantic ambiguity of the term "sharing" itself. Obviously, sharing guilt is not like sharing a cab, and even less like sharing a cake cut in equal pieces. Applied to atmospheres, this notion is largely dependent on their "soft" or "hard" conception. Indeed, whilst the soft idea states that a shared atmosphere may result from overlapping private feelings of individuals that are experiencing a similar situation and are mutually aware of this sharing (this covers, in a certain sense, the moderate, weak, and strong sharing established by Mikko Salmela),[19] the hard one asserts that what is collectively experienced is a numerically single feeling. Both options reject the individualism and introjectionism that compromise most approaches to feelings, but do so starting from very different ontological-phenomenological assumptions.

The second issue is the notion of "intentional." Whilst today's debate on collective emotions doesn't give up this concept, a neo-phenomenological atmospherology considers it overestimated and misleading. What appears to be affectively intentional could be actually reinterpreted, according to Schmitz, in the terms of Gestalt psychology, namely by distinguishing within feelings a sphere of condensation, where their characteristic features are gathered, and an anchoring point, from which they issue forth as a Gestalt.[20]

The third problem is that of the spreading degree of a shared feeling. Many scholars underline that having collective feelings does not mean attributing them to a group personhood and that implying a corporate emotional and moral sensibility does not mean considering group persons as embodied agents.[21] This *caveat* applies all the more to atmospheres as felt-bodily resonating feelings. On the other hand, if a shared (atmospheric) feeling is realised for one group member if and only if it's realised among all group members, there would be no shared atmospheres. They certainly are a "we" feeling, but not in the strictly quantitative sense[22] of shared intentions.

The fourth issue is perhaps the mother of all these doubts. Whilst it's difficult to find someone openly denying the very idea of sharing feelings (or atmospheres), of a group of individuals interpreting their feelings as anonymous and apersonal rather than as their own private ones, it is significantly harder to take a stance regarding Hans Bernhard Schmid's intriguing idea of the numerical identity of certain feelings. It is true that Schmid tries to mitigate the idea that "shared feelings are conscious experiences whose subjective aspect is not singular ('for me') but plural ('for us')" (Schmid 2014, 9) through the caveat that "the numerical identity of the feeling does not preclude difference, but the difference here is one between aspects of one feeling rather than one between numerically different feelings" (Schmid 2009, 82). Nevertheless, in my view, the identity of an atmospheric feeling should be considered as a type-identity rather than as a token-identity.[23]

Schmid's ambitious double-sided approach to different feelings as aspects of one shared feeling would also be convincing, was it not for the "numerical identity of the feeling itself." Objections of various nature come undoubtedly to mind. Firstly, sharing an atmosphere might end up modifying it: sometimes reducing its qualitative impact (a sorrow shared is a sorrow halved) and other times strengthening it (a joy shared is often a joy doubled), and this fact undermines (or at least modifies) the alleged initial numerical identity (cf. Konzelman 2009). Secondly, talking about a "numerically identical atmosphere" would lump too much feelings together with "objects" or "meanings." A shared atmosphere detached from the quality, intensity, and effect of its perception sounds actually too much like Fregean "reference": but a feeling is totally different from the truth value of a sentence.

I know that these two minor objections only touch on Schmid's suggestion. So, in order to defend the argument that people only live a form of I-atmospheric feeling, I must now go into a little more detail regarding atmospheres. A full-fledged (for me prototypic) atmosphere is an initial phenomenal-affective state of indifferentiation,[24] i.e., an in-between conceived not as the result of an interaction of fixed and stable components (the self and the environment) but as a relation in a sense preceding the relata.[25] Concerning the differentiation of feelings, one could follow Schmitz[26] and drastically argue that the not perfectly identical atmospheric feeling in two different people does not prove the subjectivity and privacy of the atmosphere; or at least no more than the fact that one shelters oneself with an umbrella and another lets himself be hit by the rain does not prove the subjectivity of that.

However, instead of embracing such radical objectification, by the way often very healthy precisely because of its polemic scope,[27] I would just prefer the phenomenological model of the *epoché* to that of the hermeneutical circle. Which amounts to say that a shared atmospheric-external feeling is a spatialised feeling whose type (not token) precedes the emergence of self and other and only defines a limited horizon within the obviously much wider emotional life of different people.[28] As a real, "we" experience it acts as a framework, without forbidding relatively different token-atmospheric feelings and, as a consequence, different ontologically individuating subjective facts.[29] After all, it makes no sense to speak of sharing if this does not involve different persons.[30] The only risk I see in the idea that different felt-bodily dispositions simply filter a certain atmosphere in different ways is suggesting that there are countless coexisting atmospheres that float in the same "air" without being properly rooted in specific situational affordances. This seems to justify those critics of neo-phenomenological atmospherology, according to which it would be an undue hypostatisation and reification of the affective realm, but also, as an emotionalist stance, an irresponsible arousalism that would end up reducing people to nothing more than "blind passengers" of free-floating atmospheres. However, the risk no longer applies when stating that this first-person filtering takes place not among many atmospheric feelings but within a predominant atmosphere that can be defined as the "same" exactly for this reason.

Fortunately, or unfortunately, the issue of sharing atmospheres is the sort of matter in which examples have nothing to envy to theoretical argumentations. I, therefore, would like to give some problematic examples. Sharing a numerically identical object like a cab doesn't exclude that its atmosphere could be felt in a relatively different way by different individuals, for example, depending on whether they are regular users or not, are reimbursed or not, etc. Similarly, sharing an apartment doesn't mean that the roommates inhabit all its interiors at the same time and live the same feelings. The form of sharing exemplified here, which is material and relatively more external to persons' felt body, is something only conventional and has little to do with sharing a feeling, which must always include the felt-bodily and relatively different fine-grained qualities that one experiences.

Nor is it enough to constitute a coupled system. Even in the extreme example of a rapist and their victim, they certainly live within a shared atmosphere but do not have a numerically identical feeling, being the former involved by sexual domination and the latter by terror (as happens even in the consensual context of a sadomasochist couple). As is well known, Christmas is an influential atmospheric field[31] able to trigger different reactions among different individuals: excitement among the children waiting for gifts but also nervousness and even depression for adults forced, for example, to extenuating family reunions, excessive lunches, etc.[32] Children's joy can certainly be co-felt by adults, but only through a minor participation that is also witnessed, in fact, by the uncertainty of their felt-bodily behaviour.

Consider also how September 11 elicited (and reactivates each time) in American society an atmosphere of sadness and anger, while it aroused among al-Qa'ida soldiers an atmosphere of joy and pride. One of my favourite examples refers to an imposing entrance hall's atmosphere of a major banking institution that arouses in one fear and in another proud. What generates in these cases both feelings is precisely the "same" (but not numerical identical) spatial-emotional atmosphere, that is, a vastness and solemnity (also based on the architectural qualities). Another example is that of four components of a successful musical performance like the composer, the stage manager, the musician, and a member of the audience: they always share a collective joyous atmosphere in a somewhat different manner, according, for example, to their concerns and role differentiation.[33] Even Scheler's famous example of the parents' shared grief beside their dead child is not obvious at all.

> Two parents stand beside the dead body of a beloved child. They feel in common the 'same' sorrow, the 'same' anguish. It is not that A feels this sorrow and B feels it also, and moreover that they both know that they are feeling it. No, it is a feeling-in-common. A's sorrow is in no

way an 'external' matter for B here, as it is e.g., for their friend, C, who joins them and commiserates 'with them' or 'upon their sorrow.' On the contrary, they feel it together, in the sense that they feel and experience in common, not only the same value-situation, but also the same keenness of emotion in regard to it. The sorrow, as value content, and the grief, as characterizing the functional relation thereto, are here one and identical.

(Scheler 1954, 12 f.)

Despite the (synchronic) bodily and (diachronic) narrative intimacy strategically utilised to show their mutual attunement, made of marital love and life, biological relation and parental love, even physical closeness,[34] it cannot be ruled out that their sorrow also depends on their empathising, or even trying to reduce the other's sorrow and probably "lean on" the other. Furthermore, other factors such as a weaker commitment to caregiving, their loving each other less than before, their having experienced and known of death in different times and ways, etc., might further increase the distinction between the shared (type-feeling) atmosphere of grief and their atmospheric sorrow (token-feeling).

But things can perhaps be even more complicated. Indeed, one can certainly imagine that an atmosphere can occupy a certain space (and be third-personally perceived) without being really shared neither from those who radiate it, nor from those who perceive it, and is, therefore, a very intriguing kind of diffused feeling.[35] It can also be shared by those who radiate it but not by the perceivers (for example, focusing on something else), as well as being shared by the perceivers but not by those who are radiating it, as in the emblematic case of vicarious shame or unrequited love,[36] or, finally, be shared by all of them, as in the case of the deliberate (and successful) creation of a cheerful party atmosphere.

Nevertheless, I put these further complications aside and aim at merely expressing my view. The quasi-objective atmospheric feeling that is situationally encountered and shared (at least in a homogenous culture)[37] affects those experiencing it as a paradigmatic prefiguration, as "the atmospheric" or the "fundamental tone" (Böhme 2001, 59–71) that paves the way to all the idiosyncratic felt-bodily resonances it triggers among subjects. This leads to a distributive sharing condensed each time in objective-subjective hybrid instantiations. In other words, the same type-atmosphere[38] extends across multiple individuals, who felt-bodily live relatively different token-atmospheric experiences, an integral-qualitative part of which might also result from the fact that they are unthematically affected by the relation they have with each other.[39] The atmospheric "we" experience is nothing else than a well-balanced condition of similarity and difference.

What really makes an atmosphere a shared feeling is not its completely shared resonance but its quasi-objective-external nature. The problem is whether this necessarily implies the co-presence (physical or at least digital) of individuals to each other in the same (lived) space. The undoubted fact that people can automatically-unconsciously mimic[40] the feelings and even diseases[41] of others, and synchronise with them, easily explains why atmospherological studies are often wrongly confused with those about contagion and sometimes are tempted to even predict the social spreading of atmospheres. However, a comprehensive explanation of shared atmospheres by putting individuals first will necessarily fail. It is not enough to invoke a "contagion" (whatever that means) and it cannot simply be said that "emotions can be infectious—they can be 'caught' like colds" (Goldie 2009, 189), thus wrongly understanding contagion in the medical sense.

In fact, the person who is atmospherically infected might even be alone (where does then the contagion come from?). The atmosphere of a building or a landscape may be collectively felt, of course, and thus have a special qualitative nuance, even when we feel it alone, i.e., without necessarily having others as co-perceivers and/or as perceived. Furthermore, while the contagion is supposed to be transcultural, an atmosphere is mostly involving within limited socio-cultural configurations,[42] and above all the contagious effect can hardly be seen as a feeling that individuals have of their own (in the proper sense).[43] This seems to make the contagion theory and the over-inflated hypothesis of mirror neurons entirely superfluous, but leaves open the more metaphysical issue of whether a shared atmosphere necessarily implies that people think-imagine what the others could feel in the same situation.[44]

Since an atmosphere is a feeling neither constituted through the experience of co-attenders,[45] nor resulting from the mere summation, aggregation, and even co-regulation,[46] it can only partly be explained through empathy, which keeps a difference between self- and other-experience alive and that some may even use to deceive others,[47] or through sympathy, since I can sympathise with somebody else's atmosphere without personally sharing their feeling: for instance I feel (vicarious) shame for someone who "should" be embarrassed but is actually not and, therefore, is not experiencing the same feeling.[48]

## Atmospherological-neophenomenological suggestions

Of course, accepting the idea that a shared atmosphere is only a type-shared feeling means embracing the neo-phenomenological distinction between the atmospheric feeling as such and the affective involvement it triggers.[49] This distinction always raises a deluge of criticisms for its alleged reification, but I believe it is continuously confirmed by our everyday experience of atmospheres that we cannot resist (or change) and are even ashamed of not being able to resist (or change). In fact, atmospheres can be emotionally authoritative and binding both in a direct-dystonic form, as when one in a good mood feels oppressed by a grey and sultry morning and easily imagines that everyone else would similarly be oppressed. But they can also be binding, in a more intriguing case, in an inversed form, as when a beautiful landscape, exactly because of its beauty, sharpens someone's (previous) sadness. Being sure of feeling something not strictly personal but rooted instead in the objective situation, individuals consider the experienced atmosphere as a weak or moderate collective feeling that is different from their subjective mood and that for this reason exerts an authority over them, whether they "decide" to adapt to it or resist it.

This atmospheric/atmosphere detachability—to use Gernot Böhme's distinction between the atmosphere before it involves us and once it touches us directly (atmospheric/atmosphere)—is further proved by the fact that individuals can recognise and describe an atmosphere, thus adopting an observing point of view, without personally feeling themselves involved, i.e., without an experiencing point of view. As happens to Rainer Maria Rilke's Malte—"the people stopped me and laughed, and I felt that I ought to laugh too but I couldn't" (Rilke 2016, 29)—and to Charles Dickens' Mrs Gradgrind—"'Are you in pain, dear mother? I think there's a pain somewhere in the room,' said Mrs Gradgrind, 'but I couldn't positively say that I have got it'" (Dickens 1854, 234). Given that only a low-intensity and incipient sharing of other people's atmospheric feeling[50] can really explain and justify this kind of uninvolving perception,[51] it is an excellent example of individuals recognising and, therefore, weakly sharing a feeling without being fully involved in it. This recognising without really sharing obviously happens in an involuntary way, because if individuals could always merely observe the environing feelings, they would have an infallible tool to neutralise and even get rid of the undesired ones.[52]

That an atmosphere is shared means that its affective widespread quality identifies and distinguishes a situation from the other,[53] leading to what is salient or not, possible or not. That is why every discussion on collective atmospheres also requires an in-depth examination of the possible types of situations. Using Hermann Schmitz's classification, it could be said that a situation is present or long-standing, impressive or segmented, deeply rooted or only inclusive, and that there are, as a consequence, different forms of atmosphere sharing, whose binding power depends both on the felt-bodily disposition of the people involved each time as well as on the degree of the atmosphere's situational rootedness. To give just an example: there are atmospheric feelings remaining the same even when they are rejected and not shared (a landscape, for example, may be melancholic as such even if the spectator is happy), and other, more resonance-conditioned, existing only when they are embodied and shared (without brave people, for example, there can be no proper atmosphere of courage).

## Collective affective states

We must bear in mind, however, that current language often uses "atmosphere" to also mean a socio-historical collective mood. Joseph De Rivera (1992), for example, suggests distinguishing between three types of collective affective states. For him they are objective in a certain sense, that is, insofar as they provide an identity to society and govern the interaction between its members. They are transitory emotional atmospheres as short-term, situation-related affective group experiences focused on a particular event; intersubjective-social emotional climates reflecting longer term socio-political conditions; and broad affective cultures.

This sociological differentiation should not be underestimated just because it is rather too simple. It reminds me of my own inflationary distinction, based on increasing objectivity and authority, between spurious, derivative-relational, and prototypic atmospheres, albeit with some differences I intend to focus on. Firstly, I call atmospheres the long-lasting, more pervasive and unthematic states that De Rivera instead calls climate and culture (derivative atmospheres), but also those resulting from a subjective projection (spurious atmospheres) and above all the feelings that are independent of humans and come out of the blue (prototypic atmospheres). Secondly, people can be grasped by (and share) atmospheres even where they are much less aware of them than De Rivera assumes.[54] After all, it is well known that numbness, vagueness, and twilightness, precisely by precluding a clear separation from the environment, seem to strongly improve the atmospheric experience.[55]

The very fact that De Rivera's "emotional climate" is not really independent of local-temporary group atmospheres and from historical-national affective culture clearly demonstrates that the term "atmosphere" can legitimately cover the whole field of collective feelings. In my view, it also does justice to what Schmitz calls "historic climate" and "style," referring to the sedimentation of pre-atmospheric felt-bodily basic moods that, in turn, help selecting what can in each moment atmospherically build a culture.[56] The surprising correspondence between the brilliant trumpet sound in baroque music and an agile felt-bodily disposition, for example, confirms for Schmitz exactly such a felt-bodily-shared atmospheric culture.

Nevertheless, here doubts emerge again. What is now, for example, the Italian atmosphere today if meant as a historic climate? At first glance, it is easy to answer by talking about instability, distrust of institutions, populism, mistrust of the future, paranoid need of identity, etc. But it also certainly applies to other countries and it's difficult, of course, to conclusively establish how many individuals (including maybe imaginary ones) and what time frame a collective atmosphere has to involve. Aggregate macro-studies do not secure a successful outcome, since people responding to questionnaires might be afraid or ashamed of saying what they really feel, or simply say what they believe the others want to hear. The risk is, furthermore, that of adopting different standards of comparison, confusing aspirations (what people would like to happen), entitlements (what has to happen), and observations (what happens), temporary and more lasting moods, personal or even strictly characterial feelings and collective ones.

Needless to say then that quantitative studies, as well as being philosophically insufficient, often underestimate that people of different neighbourhoods, regions, social classes, families, ages, etc., may perceive their time quite differently, and that it is far from being clear how a collective atmosphere is temporally and geographically extended.[57] Do replies to a questionnaire reflect the way individuals feel their environment or simply how they think the majority of the others are feeling? Is the really shared atmosphere of a community the feeling it prefers (a quiet never-changing life, for example), or that which it dreams (a more innovative life)? Is it what people feel at home or what they feel, by contrast, when they are talking with strangers or live temporarily abroad and unexpectedly realise (and appreciate) the state of well-being they unwittingly enjoyed at home? Or, again, is it a feeling in act (the distrust of the current government, for example) or the long-standing one (the confidence in the democratic reliability of their own country)? Lastly—but the list could go on, of course—even the feeling of being lonely and distrustful of any collective dimension can be shared. What one experiences here is, in a paradoxical but philosophically evocative way, an atmosphere that is shared but made of not shared feelings, which, by the way, can sometimes even reinforce the social cohesion and positive affect within a smaller social group.[58]

## Atmosphere management

What about politics then? A society unable to support the necessary (positive or at least negative) shared atmosphere certainly disintegrates, but there is an obvious difference between an authoritarian-vertical integration (patriarchal authoritarian atmosphere) and a more democratic-horizontal one (fratriarchal atmosphere).[59] In any case, every integration aimed at maintaining, refreshing, and strengthening social bonds, of course, is not without costs,[60] and unfortunately De Rivera's appeal to "caring communities that are open to others" (2014, 229)[61] is too generic to be politically useful.

The key issue of the ethical responsibility and political management of atmospheres merits a deeper debate and it is not to be pursued further here.[62] After all, it is not even so sure that governments' policies have more responsibility for emotional life of individuals than the expressive qualities of their environment, or that people are really able to intentionally generate atmospheres. I clarified elsewhere[63] why one can legitimately get suspicious about affective-atmospheric voluntary planning. In fact, designing atmospheres intentionally (i) might just produce a sentimentalist and naïve (kitsch) situation;[64] (ii) maybe it is impossible, because atmospheres belong to a preconscious sphere that is prior to the subject/world dualism;[65] finally (iii) it might lead to an instrumental-unethical and only demagogical kind of emotional hygiene.[66]

Nevertheless, daily experience tells us something less ethically dangerous. When, for example, I go back to home and, after lowering the lights and closing my blinds, I let myself go and enjoy my favourite music, I am certainly able to generate the desired atmosphere through a specific affective-environmental niche.[67] Just in planning a certain environment, made of specific things (a book, an armchair, a cup of tea, alone or with friends), practices, not only precognitive expectations, but above all of a very particular light (dimmed, diffused, warm), consists what the Danes call *hygge*. This term, which has almost become a national symbol with double value (descriptive and normative), means a situation of security, familiarity, intimacy, relaxation, generally well-being experienced at the moment thanks to the temporary interruption of the normally anxiogenic time flow.[68] If no atmosphere was voluntarily designed, no Dane could ever experience the one defined as *hygge*, nor could a work of art or a spectacle, a rally or a popular demonstration never be atmospheric effective—which is obviously absurd. Maybe what can be said is (and it's quite different) that one can stage only derivative and spurious atmospheres, not the prototypic ones. It is clear, for example, that I cannot, just desiring it, generate the Christmas atmosphere in the heat wave of August and without what it entails (the narrative, social, and environmental setting made up of decorations, lights, cold, Santa Claus, perhaps snow, etc.).

In addition, although in general an intentional atmospheres management is possible, as Gernot Böhme tells us rightly talking about aesthetic workers (in a broad sense) and generators of atmosphere like characters, synaesthesia, social characters, etc.,[69] it is always made difficult by their lingering nature. Unlike a rapidly vanishing state like emotions, the atmosphere of a person and even a society can eventually maintain and strengthen, or conversely reduce and even lose its power due to many factors.[70] As soon as we introduce the time variable, speaking of long-term shared atmospheric feelings becomes more difficult. Atmospheric feelings, in fact, are interwoven with one another over time, reveal possibly conflicting sub-atmospheres within them, etc.[71]

Let's now return to the atmosphere of the bank we mentioned above. The initial shyness and anxiety of individuals hoping for a loan might, for example, turn into pride (of being among those the bank can trust), compassion (for those, instead, the bank does not help), esteem (for employees more available than expected), admiration (for the advanced digitalisation of financial transactions), satisfaction (for obtaining the loan), etc., or at least merge with them. Given this time variable, one simply doesn't know any more if the shared atmosphere is that triggered by the first impression of environmental expressive qualities and affordances and that paradigmatically seeps in the subsequent life of the involved person[72] (this may seem minor, but, however great the power of a projective-constructionist feeling, a major banking institution cannot atmospherically be what it really is if it were a shack or a small and poor apartment on the second floor...). Or if the shared atmosphere is rather the following one, arising out of a changing mood or a blend of not always coherent feelings. What is important for my atmospherology is that even when individuals reject or hybridise the feeling elicited by the first impression, they are still rejecting or hybridising the initially felt and type-shared atmosphere.

At this stage, one could also raise the objection of the so-called affect theorists. They would claim that a shared atmospheric feeling is neither a feeling nor an emotion but rather an affect, meaning by this a pre-personal and non-conscious, formless and pre-linguistic, meaningless and corporeal potential of intensity. In truth, a neo-phenomenological atmospherology has no trouble defining the felt-bodily resonance to atmosphere as a form of sub-personal bodily thinking and explaining it, according to Schmitz's felt body "economy" and "alphabet," as an endless modification of the contraction-expansion relationship. And it would also have no trouble to consider atmospheres irreducible to the physical-bodily conscience and to the post-hoc rumination (maybe also conventional and ideological) of mind and language. However, this does not necessarily mean that an atmosphere is an unpredictable and indeterminate cosmic entity that is hardly distinguishable from matter movements and involves individuals without any regard to its content.[73] In political terms, which are notoriously the most burning consequences of new affect theories, an atmosphere contains its own meanings and expresses them through its immanent affordances.

A good political atmosphere, therefore, depends not just on personal taste and/or a certain degree of intensity, but rather appeals to quasi-normative criteria that should be able to limit, as far as possible, toxic and manipulative atmospheres. In short, one cannot be content with simply opposing to some manipulated atmospheres other equally manipulated ones, but they must also demand an atmospheric "competence."

An atmospheric competence should be able, first of all, to distinguish between "toxic" and "benign" atmospheres. It should also be able to accept the fact that, due to the lack in our post-traditional societies of a paradigmatic place of atmospheric awareness—that is, of a situation that may act as a paradigm of every other atmospheric experience—one should rather learn to have as many different atmospheres as possible and thus to allow the resulting experiences to interact with one another. This could give rise to a well-being that, exactly as happens in democracy, depends on a division of powers (affective in this case) that relativises their impact in a beneficial way. Lastly, a good atmospheric competence should favour and foster those atmospheres where, as happens with a *trompe l'oeil*, an early pathic-immersive step may and should be followed by an emersion phase.

This combination of immersion and emersion, involvement, and distance[74] happens especially in contemporary art,[75] which, unlike populism, in fact, generates cognitive and affective discontinuities through its provocatory and irritating impact. And these discontinuities always create a critical distance as well as empower whoever deeply experiences them. I do not see other ways of immunising oneself (always only partially) against today's pervasive atmospherization. The usual ways proposed to do so are romantic-moralistic or cognitive-naturalistic, but they are naive either way. The former kind overestimates personal freedom and the critical power of merely cognitive scepticism; the latter is unrealistically ascetic in requiring a big distance from the affective world, thus degrading human beings to neutral observers of life: a view from nowhere that, as a lack of felt-bodily resonance, could also become a symptom of a psychopathological crisis. My pathic aesthetics, on the contrary, "subversively" promotes an ontological and phenomenological inflationism that even includes quasi-things, and, being moderately critical towards today's predominant stage-values, limits itself to proposing a "provisional atmospheric morality" (to paraphrase Descartes).

## Conclusion

I certainly raised here more (also political, sociological, etc.) issues than I could ever solve. In order to avoid being surprised by the not perfectly rational behaviour of communities, I would like to highlight the widespread influence of atmospheres on the private and collective climate. My differentiation between three kinds of atmosphere, furthermore, aims at accounting for the extent of atmospheric experience and at identifying ongoing but (ontologically and phenomenologically) different transactions with the world. These three kinds also imply different types and grades of emotional sharing, from the most shared prototypic atmospheres down to idiosyncratic spurious ones, as well as the temporary dominance of one of the two felt-bodily poles, in a continuum that goes from the more external atmosphere down to the projective-subjective one.

But the way in which atmospheric feelings can be shared through felt-bodily processes is still an open issue and speaking of "sensitive communication and certainty"[76] is just a first step. Not only because the paradoxical but far from impossible situation in which "I feel something I don't share" should be explained, but also because an entirely shared atmosphere seems a pretty rare experience within the logic of felt-bodily communication:[77] only one of its forms, the solidaristic incorporation, makes sharing and reciprocity possible through a collective type-shared atmosphere generating a collective felt body. All other forms of felt-bodily communication actually imply a felt-bodily understanding of the others and a coordinate coupling that implies, just like dance and sport, a shared structure equality but also partially or even fully different felt-bodily feelings.

In conclusion, my angle on the issue is that one can feel the same type-atmosphere but not really feel what the token-atmospheric feeling is like for others. Needless to say, that a detailed (for example, socio-statistic) analysis designed to remove the structural vagueness of atmospheres seems here altogether out-of-place. Being, for example, Italian, of the same age, fellow citizens, etc., is not enough to understand and predict the felt-bodily resonance of which a spatial we-feeling is capable.[78] Nor can this sharing be proved by directly asking people about their atmospheric feeling, since this could be too subtle and nuanced to be put in words, or, as is well known, they would likely give not fully sincere answers or simply repeat commonplaces. After all, every too analytic-quantitative enquiry of our involuntary atmospheric experience is, therefore, doomed to fail, if only for the fact that the scientific method is—said with the necessary cynicism—"the institutionalized maintenance of sangfroid in the face of surprise" (Massumi 2002, 233). However, is an experience without surprises still a real experience?

## Notes

1 Rightly emphasising their unconscious, automatic, and involuntary aspects, in fact, does not mean (contrary to what Krueger 2014, 158 says) that feelings are necessarily intrapsychic-individual states.
2 This not only in early infancy (as suggested by Krueger 2013, 2014, developing Merleau-Ponty's idea), because if that were the case, only in childhood one would really experience atmospheric feelings.
3 A view that is ontological (feeling are necessarily feelings of somebody), epistemological (only individuals have a first-person authority regarding their affective life), and physical (see Schmid 2009).
4 Introducing briefly my conception of "atmosphere," it could be said that, never wholly detached from its climatic meaning of immersion in the weather-world, the term "atmosphere" means the "something more" one feels (senses, perceives...) "in the air," namely in a certain space or situation. It is, more specifically, a feeling poured out into a certain (pre-dimensional) space, inextricably linked to felt-bodily processes and characterised by a qualitative microgranular presentness, that is, by definition, inaccessible to a naturalistic-epistemic perspective. In this sense, an atmosphere is also an example of the passive synthesis largely intersubjective and holistic that precedes analysis and influences from the outset the emotional situation of the perceiver, resisting mostly any conscious attempt at projective adaptation or epistemic correction.
5 Cf. Schmitz (1969, 1992, 2007, 2014, 2019) and Böhme (1995, 1998, 2001, 2006, 2017a, 2017b).
6 Provided, however, that "atmosphere" is not only a misleading metaphor (as Schmid 2014, 10 claims). This downplays its necessary felt-bodily nature and the fact that something that does not have a parallel and fully equivalent literal sphere cannot, strictly speaking, be called metaphorical.
7 Maybe, it is impossible to describe a feeling in non-spatial terms. Atmospheres, however, only refer to the "absolute" (lived, pre-dimensional, anisotropic) space to which subjects participate thanks to their felt body, i.e., their facing "pathically" cross-modal "ecological" suggestions. An atmosphere (especially if prototypic) has among its characteristics to have authority (in principle) and try to occupy a certain space (see Griffero 2014b).
8 This is certainly true, taking here up a Schmitz's contentious distinction, more for the widely spread atmospheres of feeling than for the just body-related ones.
9 An atmosphere, in fact, doesn't usually function (unlike emotions) as a "fast track [...] to ownership, mental unity, relevance and commitment" (Schmid 2014, 8).
10 As is known since Hubertus Tellenbach's seminal book on atmosphere (Tellenbach 1968).

11 For an introduction to the notion of felt-bodily "resonance" (cf. Griffero 2016, 2017c, 2020b).

12 Cf. Krueger 2014.

13 Even if Schmitz (1999, 280) considers it reductive to describe New Phenomenology as a philosophy of felt body (*Leiblichkeit*), he makes a real Copernican revolution when he aims at conceiving affective life and life *tout court* in a new (felt-bodily involving) experiential way.

14 This category (*Halbding*) first appeared in Schmitz (1978, 116–139) and was partially inspired by Sartre's pages on pain as a psychic-affective object with its own reality, intermittent time and life, habits and "melodic" developments (Sartre 1978, 335–337). By saying that quasi-things occupy a vast territory between mere *qualia* and full-fledged things one gives due value to all that, despite its vagueness, transience, fluidity, and lack of borders, they (unwillingly) experience very frequently. Think, for example, of the wind, gazes, sounds, colours, the night, certain thermal qualities, smells, electric shocks, weight and void, time, and, above all, atmospheric feelings. These are sets of *qualia* that, for their marked expressiveness, physiognomic "character" and intrusiveness, affect people more than full-fledged things. Quasi-things are usually dismissed in two manners: a) by forcedly "thickening" and turning them into things in order to reduce their particular intrusiveness: for example, by reducing the wind to air in motion when it blows or still when it dies down; b) by tracing them back to perceptions so chaotic that they end up being considered as something anomalous, if not pathological. On the contrary, quasi-things are for me a salient and primary component of the passive synthesis through which something happens to us and felt-bodily involves us. See Griffero (especially 2017a).

15 Social emotion-externalism is promoted, for example, by Léon, Szanto, Zahavi (2017).

16 See Schmitz (2011, 29–53) and, for a general survey on the debate on *Stimmungen* (or moods) from an atmospherological point of view, see Griffero (2019b).

17 It is obvious that "grief is not localized in the body, as if one's grief were felt in one's left leg" (Schmid 2009, 72). However, it certainly resonates (as heaviness or the haze, for example) in the felt body.

18 "But being affected by the unserviceable, resistant, and threatening character of things at hand is ontologically possible only because being-in as such is existentially determined beforehand in such a way that what it encounters in the world can matter to it in this way. This mattering to it is grounded in attunement, and as attunement it has disclosed the world, for example, as something by which it can be threatened. Only something which is the attunement of fearing, or fearlessness, can discover things at hand in the surrounding world as being threatening. The moodedness of attunement constitutes existentially the openness to world of Da-sein." (Heidegger 1996, 129).

19 Following Salmela, in fact, we have three options: a) a moderate sharing, when individuals affectively evaluate an event similarly from the perspective of their identitary shared concern (the joy for a goal scored by own team), b) a weak sharing (the shared atmosphere of solemnity of a certain building), and c) a strong sharing, based on an individuals' collective commitment to a concern as a group (the atmosphere of victory of players of a sport team).

20 This means that sometimes the intentional object is only apparent, because the centring may not be found in the anchoring point of the percept (its generative point) but in its sphere of condensation (where the character of the percept is concentrated), so that it is not unusual to find the anchoring point "at the opposite end of the condensation zone" (Metzger 1941, 230). Sometimes a

lover loves (atmospherically) even the doormat of their beloved's house; this doormat, surrounded by a positive atmosphere, is obviously only part of the sphere of condensation of love, yet is sometimes more intensely felt than the concrete presence of the beloved person (who is the anchoring point). Just as one usually does not look at the sun directly but only at the things enlightened by it, one often confuses the (perhaps occasional) condensation point of an atmospherically diffused feeling with its intentional object (Schmitz 1969, 322).

21  Cf. Szanto (2016, 270).
22  See Salmela (2012, 8) and Sánchez Guerrero (2011).
23  For this distinction see Schmid (2009, 69, 77).
24  As suggested by Krueger (2013, 2014) but also by Scheler, convinced that infectiousness atrophises when one gets older. If that were the case, one would have a real atmospheric experience only in childhood (which is simply false).
25  What Schmid (2005, 138) should perhaps admit, given his idea of an emotional sharing preceding any distinction between the experiencers.
26  Schmitz (1969, 216).
27  After all, it is precisely the unusual philosophical ideas that shed some light on why we conceive of matters in the way we normally do (Schmid 2009, 88).
28  Schmid (2009, 80) would define it an only partial fusion.
29  As Slaby (2014, 34) rightly claims in a precisely (although perhaps unintended) neo-phenomenological way: "take a person's emotionality away, and there's nothing left that deserves to be called 'self'." This also applies actually to certain involving thoughts that precisely for this reason should be considered more effective than cognitive thoughts.
30  Schmid himself (2009, 80; 2014, 12) admits a dependence on a wide array of circumstances.
31  For De Rivera (2014, 236–238), even the prison's emotional climate, arousing obviously different feelings in guards and prisoners, would be their common climate!
32  How many Hollywood movies have we seen in which family anniversaries become tragedies or at least traumatic showdowns?
33  Salmela (2012, 6).
34  See Krueger (2016, 270 ff.).
35  Cf. Schmitz (1969, 137).
36  Goldie (2009, 193–194).
37  An objection that is not irrelevant, I know, and should be discussed more thoroughly.
38  Which would correspond (in hermeneutics) to meaning and not significance (contra Sánchez Guerrero 2011, 267).
39  See Zahavi (2014, 117). As regards the feelings we deeply share with persons we haven't, however, seen in a long time are largely an illusion due to the nostalgic prominence of a very short part of our past life.
40  Through facial, vocal, postural mimicry and related feedback (Hatfield et al. 2014).
41  For example: fear, panic, joy, mania, anger, depression, embarrassment, patriotism, hatred, political and religious enthusiasm, but also allergies, obesity, smoking, sleep problems, etc.
42  León et al. (2017, 4, fn.3). It is the context of "shared cognitive and conative attitudes" (Schmid 2009, 80).
43  Zahavi (2015).
44  But can you really imagine, for example, the atmosphere of a kabuki show for Japanese as you can the rock or jazz concert one for Westerners?

45 Even if the co-presence certainly makes it harder to objectify this kind of atmosphere (Böhme 2007, 281).

46 Nor shall it suffice to call for the simple joint (moral, ritual, normative) commitment (Gilbert 2014, 23).

47 For example: empathically understanding that you love your wife is quite different from loving her myself! (Zahavi 2014, 150).

48 On vicarious shame as a quasi-thingly atmosphere see Griffero (2017a, 79–92).

49 Schmitz (1969, 87, 96).

50 Which does not necessarily imply normative (postulating that others follow social-ruled thought and feelings) or naturalistic assumptions (postulating that others' affective states are fully subject to natural laws).

51 "When we fight against penetration of an objective feeling, it has already taken hold of us" (Baensch 1924, 9).

52 As Schmitz (1969, 150), despite certain hesitations, claims.

53 Unlike Schmitz thinks (2014, 50, 55).

54 Schmid (2014, 6) speaks of pre-reflective self-awareness.

55 On atmospheres' (maybe) essential twilightness, see Griffero (2017a, 103–112).

56 See Schmitz (1966; 1992, 317–331). On the not always clear relationship in Schmitz between atmosphere and collective basic mood see Trčka (2011, 191).

57 Pierre Bourdieu's more socio-political notion of "habitus" and John Searle's more mental-conventionalist one of background are here of little use.

58 Rimé (2007, 310) rightly also remind us that sometimes social sharing interactions (in the case, for example, of severe illness) do not bring interactants closer to one another at all.

59 See Denison (1928).

60 For example: the authoritarian-vertical integration minimises the ego and sacralises obedience to authority, whilst the democratic-horizontal one maximises competition and develops totally invented and necessarily populistic emotions.

61 See also De Rivera (1992), De Rivera and Páez (2007).

62 For a few words on the subject (can democracy count on a specific atmosphere?) see Griffero (2019a, 159–166).

63 Griffero (2020c, 88–96).

64 See Geiger (1911).

65 See Heidegger (1996 and especially 1995) and Bollnow (1956).

66 See Bollnow (1956) and Schmitz (1998).

67 An environmental tool like music especially can function both as an instigator and a container of feeling (DeNora 2000, 57).

68 Mikkel Bille's research (2019) focuses on this exquisitely atmospheric notion.

69 See Böhme (2001, 2017b, 92–94).

70 Many variations in atmospheric perception depend on variations in the perceptual field (distance of the perceived, brightness of the environment, speed with which one approaches the perceived, scale shift, change of mood, unveiling of a sensory illusion, new perceiver's physiological, and felt-bodily conditions as well sensible-cognitive awareness, additional and divergent information altering affective impact, shifting of attentional focusing, purely idiosyncratic experiences, response modulation influencing the initial behavioural response, effort to minimise or change the affective impact, etc.). Not to mention the perceiver's variable sociological and/or aesthetic competence in detecting the atmospheric potential of a certain situation.

71 No matter what the poet says: "Blessed are all simple emotions, be they dark or bright! It is the lurid intermixture of the two that produces the illuminating blaze of the infernal regions" (Hawthorne 2012, 195).

72  Which one conforms to or even "capitalizes": as is well known, the memory of a good atmosphere can affect the way one appraises one's situations in the following days, weeks, etc.
73  So that when people have different affective responses, they don't disagree, they just are different (Leys 2011, 452).
74  Prütting (1995, 152) considers this "going with" or "following something" (*Mitgehen*), this combination of an understand with the belly and an understand with the head, the specific form of aesthetic pleasure.
75  Cf. Schouten (2011, 106).
76  See Gugutzer (2012, 64–67).
77  See Griffero (2017b, 2020b).
78  See Grossheim et al. (2014, 16).

# References

Baensch, Otto. 1924. "Kunst und Gefühl." *Logos* XII/1, 1–28.
Bille, Mikkel. 2019. *Homely Atmospheres and Lighting Technologies in Denmark: Living with Light*. London: Bloomsbury.
Böhme, Gernot. 1995. *Atmosphäre. Essays zur neuen Ästhetik*. Frankfurt a. M.: Suhrkamp.
Böhme, Gernot. 1998. *Anmutungen. Über das Atmosphärische*. Ostfildern v. Stuttgart: Edition Tertium.
Böhme, Gernot. 2001. *Aisthetik: Vorlesungen über Ästhetik als Allgemeine Wahrnehmungslehre*. München: Fink.
Böhme, Gernot. 2006. *Architektur und Atmosphäre*. München: Fink.
Böhme, Gernot. 2007. "Atmosphären in zwischenmenschlicher Kommunikation." In S. Debus, R. Posner (eds.), *Atmosphären im Alltag. Über ihre Erzeugung und Wirkung*. Bonn: Psychiatrie-Verlag, 281–293.
Böhme, Gernot. 2017a. *The Aesthetics of Atmospheres*. London-New York: Routledge.
Böhme, Gernot. 2017b. *Atmospheric Architectures. The Aesthetics of Felt Spaces*. London et alia: Bloomsbury.
Bollnow, Otto Friedrich. 1956. *Das Wesen der Stimmungen*. Frankfurt a.M.: Klostermann.
Denison, J. H. 1928. *Emotion as the Basis of Civilization*. New York: Charles Scribner's Sons.
DeNora, Tia. 2000. *Music in Everyday Life*. Cambridge: Cambridge University Press.
De Rivera, Joseph. 1992. "Emotional climate: Social structure and emotional dynamics." *International Review of Studies on Emotion* 2, 197–218.
De Rivera, Joseph. 2014. "Emotion and the formation of social identities." In C. von Scheve, M. Salmela (eds.), *Collective Emotions. Perspectives from Psychology, Philosophy, and Sociology*. Oxford: Oxford University Press, 217–231.
De Rivera, J. & Páez, D. 2007. "Emotional climate, human security, and cultures of peace." *Journal of Social Issues* 63, 2, 233–253.
Dickens, Charles. 1854. *Hard Times. For These Times*. London: Bradbury & Evans.
Geiger, Mortiz. 1911. "Zum problem der Stimmungseinfühlung." *Zeitschrift für Ästhetik und allgemeine Kunstwissenschaft* 6, 1–42.
Gilbert, Margaret. 2014. "How we feel: Understanding everyday collective emotion ascription." In M. Salmela, C. von Scheve (eds.), *Collective Emotions. Perspectives from Psychology, Philosophy, and Sociology*. Oxford: Oxford University Press, 17–31.

Goldie, Peter. 2009. *The Emotions. A Philosophical Exploration*. Oxford: Clarendon Press.

Griffero, Tonino. 2014a. *Atmospheres. Aesthetics of Emotional Spaces*. London-New York: Routledge.

Griffero, Tonino. 2014b. "Who's afraid of atmospheres (and of their authority)?" *Lebenswelt. Aesthetics and Philosophy of Experience* 4, 1, 193–213.

Griffero, Tonino. 2016. "Atmospheres and felt-bodily resonances." *Studi di estetica* 44, 1, 1–41.

Griffero, Tonino. 2017a. *Quasi-Things. The Paradigm of Atmospheres*. New York: Suny.

Griffero, Tonino. 2017b. "Felt-bodily communication: a neophenomenological approach to embodied affects." *Studi di estetica* 45, 2, 71–86.

Griffero, Tonino. 2017c. "Felt-bodily resonances. Towards a pathic aesthetics." *Yearbook for Eastern and Western Philosophy* 2, 149–164.

Griffero, Tonino. 2019a. *Places, Affordances, Atmospheres. A Pathic Aesthetics*. London-New York: Routledge.

Griffero, Tonino. 2019b. "In a neo-phenomenological mood. Stimmungen or atmospheres?" *Studi di estetica* 48, 2, 121–151.

Griffero, Tonino. 2020a. "Emotional atmospheres." In T. Szanto, H. Landweer (eds.), *The Routledge Handbook of Phenomenology of Emotion*. London-New York: Routledge, 262–274.

Griffero, Tonino. 2020b. "Better be in tune. Between resonance and responsivity." *Studi di estetica* 48, 2, 90–115.

Griffero, Tonino. 2020c. "Was kann eine Gefühlatmosphäre tun? Atmosphären zwischen Immersion und Emersion." In B. Wolf, C. Julmi (eds.), *Die Macht der Atmosphären*. Freiburg-München: Alber, 77–96.

Griffero, Tonino. 2021. *The Atmospheric "we". Moods and Collective Feelings*. Milan: Mimesis International.

Grossheim, M., Kluck, S. & Nörenberg, H. (2014). *Kollektive Lebensgefühle. Zur Phänomenologie von Gemeinschaften*. Rostocker Phænomenologische Manuskripte 20.

Gugutzer, Robert. 2012. *Verkörperungen des Sozialen. Neophänomenologische Grundlagen und soziologische Analysen*. Bielefeld: Transcript.

Hatfield, E., Carpenter, M. & Rapson, R. L. (2014). "Emotional contagion as a precursor to collective emotions." In C. von Scheve, M. Salmela (eds.), *Collective Emotions. Perspectives from Psychology, Philosophy, and Sociology*. Oxford: Oxford University Press, 109–122.

Hawthorne, Nathaniel. 2012. *Tales*. New York: Norton.

Heidegger, Martin. 1995. *The Fundamental Concepts of Metaphysics: World, Finitude, Solitude*. Bloomington-Indianapolis: Indiana University Press.

Heidegger, Martin. 1996. *Being and Time*. New York: Suny.

Konzelman Ziv, Anita. 2009. "The semantics of shared emotions." *Universitas Philosophica* 26, 52, 81–106.

Krueger, Joel. 2013. "Merleau-Ponty on shared emotions and the joint ownership thesis." *Continental Philosophy Review* 46, 4, 509–531.

Krueger, Joel. 2014. "Emotions and the social niche." In C. von Scheve, M. Salmela (eds.), *Collective Emotions. Perspectives from Psychology, Philosophy, and Sociology*. Oxford: Oxford University Press, 156–171.

Krueger, Joel. 2016. "The affective 'we': Self-regulation and shared emotions." In: T. Szanto, D. Moran (eds.), *Phenomenology of Sociality. Discovering the 'we'*. New York: Routledge, 263–277.

Lawrence, D. H. 1995. *The Rainbow*. Ware: Wordsworth.

León, F., Szanto, T., Zahavi, D. 2017. "Emotional sharing and the extended mind." *Synthese* 196, 4847–4867.

Leys, Ruth. 2011. "The turn to affect: A critique author(s)." *Critical Inquiry* 37, 3, 434–472.

Massumi, Brian. 2002. *Parables for the Virtual: Movement, Affect, Sensation*. Durham: Duke University Press.

Metzger, Wolfgang. 1941. *Psychologie. Die Entwicklung ihrer Grundannahmen seit der Einführung des Experiments*. Darmstadt: Steinkopf.

Prütting, L. 1995. "Über das mitgehen. Einige anmerkungen zum Phänomen transorchestraler Einleibung." In M. Großheim (ed.), *Leib und Gefühl. Beiträge zur Anthropologie*. Berlin: Akademie Verlag, 141–152.

Rilke, Rainer Marie. 2016. *The Notebook of Malte Laurids Brigge*. Oxford: Oxford University Press.

Rimé, Bernard. 2007. "The social sharing of emotion as an interface between individual and collective processes in the construction of emotional climates." *Journal of Social Issues* 63, 2, 307–322.

Salmela, Mikko. 2012. "Shared emotions." *Philosophical Explorations* 15, 1, 1–14.

Sánchez Guerrero, Hector. 2011. "Gemeinsamkeitsgefühle und Mitsorge: Anregungen zu einer alternativen Auffassung kollektiver affektiver Intentionalität." In J. Slaby, A. Stephan, H. Walter, S. Walter (eds.), *Affektive Intentionalität. Beiträge zur Welterschließenden Funktion der Meschlichen Gefühle*. Paderborn: Mentis, 252–282.

Sartre, Jean-Paul. 1978. *Being and Nothingness*. New York: Pocket Books.

Scheler, Max. 1954. *The Nature of Sympathy*. London: Routledge and Kegan Paul.

Schmid, Hans Bernhard. 2005. *Wir-Intentionalität: Kritik des ontologischen Individualismus und Rekonstruktion der Gemeinschaft*. Freiburg: Karl Alber.

Schmid, Hans Bernhard. 2009. *Plural Action. Essays in Philosophy and Social Science*. Dordrecht et alia: Springer.

Schmid, Hans Bernhard. 2014. "The feeling of being a group: Corporate emotions and collective consciousness." In C. von Scheve, M. Salmela (eds.), *Collective Emotions. Perspectives from Psychology, Philosophy, and Sociology*. Oxford: Oxford University Press, 3–16.

Schmitz, Hermann. 1966. *System der Philosophie*, II: 2, Der Leib im Spiegel der Kunst. Bonn: Bouvier.

Schmitz, Hermann. 1969. *System der Philosophie*, III: 2, Der Gefühlsraum. Bonn: Bouvier.

Schmitz, Hermann. 1978. *System der Philosophie*, III: 5, Die Wahrnehmung. Bonn: Bouvier.

Schmitz, Hermann. 1992. *Leib und Gefühl. Materialien zu einer philosophischen Therapeutik*. Paderborn: Junfermann.

Schmitz, Hermann. 1998. "Situationen und Atmosphären. Zur ästhetik und Ontologie bei Gernot Böhme." In M. Hauskeller et al. (eds.), *Naturerkenntnis und Natursein*. Frankfurt a.M.: Suhrkamp, 176–190.

Schmitz, Hermann. 1999. *Der Spielraum der Gegenwart*. Bonn: Bouvier.

Schmitz, Hermann. 2007. *Der Leib, der Raum und die Gefühle*. Bielefeld-Locarno: Sirius.

Schmitz, Hermann. 2011. *Der Leib*. Berlin-Boston: de Gruyter.

Schmitz, Hermann. 2014. *Atmosphären*. Freiburg-München: Alber.

Schmitz, Hermann. 2019. *New Phenomenology. A Brief Introduction*. Milan: Mimesis International.

Schouten, Sabine. 2011. *Sinnliches Spüren. Wahrnehmung und Erzeugung von Atmosphären im Theater*. Berlin: Theater der Zeit.

Slaby, Jan. 2014. "Emotions and the extended mind." In C. von Scheve, M. Salmela (eds.), *Collective Emotions. Perspectives from Psychology, Philosophy, and Sociology*. Oxford: Oxford University Press, 32–46.

Szanto, Thomas. 2016. "Do group persons have emotions – or should they?" In H. A. Wiltsche, S. Rinofner-Kreidl (eds.), *Analytic and Continental Philosophy: Methods and Perspectives. Proceedings of the 37th International Wittgenstein Symposium*. Berlin-New York: De Gruyter, 261–276.

Tellenbach, Hubertus. 1968. *Geschmack und Atmosphäre*. Salzburg: Müller.

Trčka, Nina. 2011. "Ein klima der Angst. Über Kollektivität und Geschichtlichkeit von Stimmungen." In K. Andermann, U. Eberlein (eds.), *Gefühle als Atmosphären. Neue Phänomenologie und philosophische Emotionstheorie*. Berlin: AkademieVerlag, 183–211.

Wiesing, Lambert. 2014. *The Philosophy of Perception. Phenomenology and Image Theory*. London-New York: Bloomsbury.

Zahavi, Dan. 2014. *Self and Other. Exploring Subjectivity, Empathy, and Shame*. Oxford: Oxford University Press.

Zahavi, Dan. 2015. "You, me, and we: The sharing of emotional experiences." *Journal of Consciousness Studies* 22, 1–2, 84–101.

# 2 Tuning the world

## A conceptual history of the term *Stimmung* part two

*Gerhard Thonhauser*

## Introduction

Despite a rich history of emotion research since Antiquity, the term *Stimmung* was only introduced into the German language during the 18th century. Moreover, the term *Stimmung* has no direct equivalent in other major European languages. Comparing it to the English language, *Stimmung* combines the semantic fields of mood, attunement, and atmosphere. *Stimmung* signifies a thought pattern that evades binary distinctions of objective and subjective, internal and external, mental and bodily. Studying the conceptual history of the term shows that *Stimmung* has undergone several semantic shifts.[1] Most importantly, those shifts have altered who or what is considered to possibly be in *Stimmung*: We will see that the term originally was not primarily applied to the minds of sentient beings, but rather to other entities as diverse as musical instruments, the nervous system, artworks, and ontologically more opaque matters like landscapes. Moreover, related notions from the same semantic field are even applicable to scientific models and engineering processes.

In an earlier paper, I reconstructed the rich conceptual history of the term from its inception in the 18th century until the beginning of the 20th century (Thonhauser 2020b). In this chapter, I summarise the main findings of this earlier paper, focusing on three stages: the initial period from the first usage via Kant's pivotal appropriation to Romanticism; the psychologisation of the term towards the end of the 19th century that came along with the establishment of psychology as a scientific discipline; and contemporaneous counter tendencies opposing the mounting dominance of psychological explanations, most importantly within art history and life philosophy.

In this chapter, I focus on how this history culminated in Heidegger's account of *Stimmung*, which draws upon earlier accounts and transforms them into a unified framework in which *Stimmungen* are understood neither as subjective mental states nor as characteristics of objects, but as serving a world-disclosing function for our being-in-the-world as a whole. Before I reconstruct the core of Heidegger's account of *Stimmung*, I review the most obvious English translation of *Stimmung*: mood, attunement,

DOI: 10.4324/9781003131298-3

and atmosphere. This will show how each of these terms captures one important aspect of the semantic field of *Stimmung* but leaves aside other aspects. In sum, this paper aims to show that the conceptual history of the term *Stimmung* contains as yet unexplored potentials for a relational and dynamic ontology.

## The history of the term prior to the 20th century

This section provides a brief summary of the history of the term *Stimmung* prior to Heidegger. The summary builds on my previous work on the conceptual history of *Stimmung* and I advise readers to refer to that previous work for longer explications and further references (Thonhauser 2020b).

### From the first usages to romanticism

The term *Stimmung* was first used in the domain of music. The verb *stimmen* means *to tune* an instrument. When an instrument is tuned, it is *gestimmt*. But it can also be untuned, which is *verstimmt*. The noun *Stimmung* is derived from the verb *stimmen* (Grimm and Grimm 1998). In its original context, *Stimmung* refers to three related aspects of musical tuning. First, it can denote the *process* of tuning. Second, it can designate the *result* of the process, the state of being tuned. Third, it can describe the *disposition* that enables an entity, in this case, a musical instrument, to be tuned (Wellbery 2003). As we will see, the English verb *to tune* is a direct equivalent to the German verb *stimmen* in the musical context, but otherwise relates to a different semantic field. The German verb *stimmen* is related to the noun *Stimme*, which designates the human voice. This term is derived from the Ancient Greek *stoma*, which means mouth (Kluge 2012).

A search in *Google Ngram Viewer* shows that whereas the verb *stimmen* (like the noun *Stimme*) can be found throughout the entire possible search range from 1500—2019, the use of the noun *Stimmung* only dates back to the second half of the 18th century. Since then, it has remained at a nearly constant level.[2] Towards the end of the 18th century, *Stimmung* quickly emerged as a term with very broad applications. I rely on the research by Caroline Welsh (2008; 2009a; 2009b) here, who has shown that during that period *Stimmung* can be found in diverse fields of knowledge such as physiology, psychology, psychiatry, and aesthetics. *Stimmung* was a lively metaphor that circulated through various fields of knowledge. Researchers made use of different aspects of the musical meaning of *Stimmung* to develop new explanatory approaches to the phenomena they were investigating. Thus, *Stimmung* did not refer to a specific phenomenon, it rather offered a thought pattern that enabled new ways of theorising and thereby created new knowledge. Generally speaking, *Stimmung* denoted a certain idea about the interplay between entities. Depending on the context, the focus could be on the *process* of tuning, the *state* resulting from that process, or the *disposition* to engage in such a process. Interestingly, unlike terms like resonance or harmony, which only refer to the process or state of being in sync, *Stimmung* can also refer to entities being out of sync (*Verstimmung*). Moreover, it allows to stress how tuning can change (*Umstimmung*). Thus, it offers a thought pattern to explain cases on the entire spectrum from felicitous to failed interactions. Finally, *Stimmung* allows to also capture the self-acting of any kind of organism or complex system, which can bring itself in and out of sync with itself or with its surrounding (*Einstimmung*). One might hypothesise that this provided a first glimpse of the idea of self-regulating dynamic systems, which was only much later theorised by cybernetics (Wiener 1948). It could also be related to the way in which bodies are considered to mutually affect each other explored by recent Affect Theory (Gregg and Seigworth 2010). Be that as it may, the core idea of the thought pattern provided by the metaphor of *Stimmung* was that the self-attunement (*Eigenstimmung*) of an entity can explain differences in its internal functioning as well as its interplay with its surrounding.

This idea of self-tuning is also at the core of Kant's aesthetic theory, which is the first major turning point in the history of the term *Stimmung*. The key question in Kant's *Critique of Judgement* (2007) is how an aesthetic judgement can be both subjective and universal. Kant claims that aesthetic judgement cannot have the same objective validity as cognitive judgement. As judgements of taste, they cannot be settled by conceptual capacities. Nevertheless, they are not just relative to the observer, in the sense that there is no accounting for taste. Instead, Kant maintains that an aesthetic judgement about something involves the claim that others ought to judge it in the same way. So, the core issue is to explain how the universality of aesthetic judgements is possible. To solve this puzzle, Kant makes use of the above-explained metaphor of *Stimmung*. Kant explains that in an aesthetic judgement, the faculties of imagination and understanding are in appropriate accord or attunement (*Stimmung*) with each other. In other words, it is the self-tuning of the faculties of the mind that enables aesthetic judgements. Fully in line with the musical metaphor, the self-tuning of the mind does not imply that those judgements are relative to each empirical subject, as the metaphor implies that it is possible to decide whether or not the mind is tuned in the right way so that it is in sync with itself. If the self-tuning of the mind is done in a proper way, the proportionality of the faculties ensures the universal validity of an aesthetic judgement, as any properly tuned mind will reach the same judgement, enabling universal communicability of those judgements.

Kant's account of an aesthetic judgement was immensely influential and established the term *Stimmung* as an aesthetic concept. However, the following history of aesthetic theory shows that the universality of aesthetic judgement became less and less important, as the idea of

*Stimmung* as a dispositional state of the mind became the core theme. Most prominently, this theme can be traced in the works of Schiller and Schopenhauer. Schiller's core idea in *On the Aesthetic Education of Man* (Schiller 1982) is almost incomprehensible without factoring in a pointer given by another peculiarity of the semantic field of *Stimmung*. The German word for determination, *Bestimmung*, is formed by adding the prefix *Be-* to the word *Stimmung*. Schiller plays with this connection when he understands the *Stimmung* of the mind as the state of being undetermined (*unbestimmt*) in the sense of being determinable (*bestimmbar*).

According to Schiller, aesthetic freedom is key in the education of human beings, and this freedom is achieved when the mind is neither solely determined by the senses nor by reason, but when both capacities are active and cancel each other out, thus leaving the mind in a state of determinability. Schiller calls this state of determinability the free *Stimmung* of the mind, which cannot be translated by a single word, but might be described as the disposition to be freely determined (Schiller 1982, 141). Schiller claims that the experience of beauty has the power to induce such a state of mind in us, thus serving a crucial educational function (Schiller 1982, 151). In comparison to Kant who emphasises the self-tuning of the mind (*Stimmung* as a process of tuning) and the resulting accord of the faculties (*Stimmung* as the state of being tuned), Schiller focuses on the determinability of the mind (*Stimmung* as a disposition to be tuned). Moreover, whereas Kant used *Stimmung* in a narrow role within his account of aesthetic judgement, for Schiller it became to designate something like a global disposition of the mind.

This last aspect is also stressed by Schopenhauer in *The World as Will and Representation* (1972 [1819]). Schopenhauer understands the aesthetic state of mind as a state of pure contemplation, in which we are no longer bound to the principle of sufficient reason. Reaching such a state of mind requires the suspension of the will. This suspension can either be reached by an external cause, or it is achievable via a self-tuning of the mind. Schopenhauer considers the latter option much more powerful, as it can induce this state of mind independent of external circumstances. Schopenhauer considers it the sign of an artistic mind to possess this ability to self-induce such a state of mind. Similar to Schiller, Schopenhauer uses the term *Stimmung* to denote this dispositional state (Schopenhauer 1972, 231). In contrast to Schiller, his focus is less on the dimension of determinability and more on the self-acting aspect of this determination (*Einstimmung* as *Selbstbestimmung*).

We can see, then, that for Kant, Schiller, and Schopenhauer the term *Stimmung* is used in the context of the mind. But contrary to today's common usage, it does not denote a specific mental state. Rather, it describes a tuning of the mind as a whole, very much in the same way as the tuning of a musical instrument.

*Psychological research*

The 19th century saw the establishment of psychology as an independent discipline. As part of that development, aesthetics became to be seen as sub-field of psychology, which led to the assumption that aesthetic experience needs to be explained within a psychological framework. In line with this development, Hermann Lotze (1868, 65) praises Kant for emphasising the subjective character of aesthetic judgement, but claims that we need to go beyond Kant in accounting for aesthetic experience in psychological terms. This stands in stark contrast to Kant's intention, as an explanation of aesthetic judgement in terms of subjective experiences leads to a relativistic understanding, according to which aesthetic judgements are a matter of individual taste. This is exactly what Kant wanted to avoid. The emerging psychological framework, however, had no room for transcendental considerations like Kant's. Within a psychological framework, it is obvious that *Stimmungen* need to be accounted for in terms of mental states. Indeed, this is so obvious that it does not require any clarification, let alone discussion (Lipps 1903; Witasek 1904; Volkelt 1905; also see the contribution by İngrid Vendrell Ferran in this volume). As a consequence, the way in which Kant, Schiller, and Schopenhauer used the term *Stimmung* has become incomprehensible within such a framework. Within psychological discourse, *Stimmung* stopped being an explanatory pattern with universal applicability. Instead, it became synonymous with the common usage of the English term *mood*.

Around 1800, the term *Stimmung* was still very much understood with reference to the tuning of instruments and the thought pattern obtained from this context allowed for innovative theorising in various fields of knowledge. Almost everything could be in *Stimmung*, as long as it was sensible to say that it can be tuned to interact with itself and its surrounding in a multiplicity of ways. Around 1900, in contrast, it was consensus among psychologists that only sentient beings could be in *Stimmung*. A consequence of this assumption is that any attribution of *Stimmung* to entities other than minds needed to be conceptualised in terms of a projection. This is the context in which the theory of empathy was developed (Lipps 1909, 223–31; Prandtl 1910; Geiger 1911). The German term for empathy, *Einfühlung*, literally means to feel something into something else. When someone experiences a meeting as tense or a foggy autumn day as gloomy, the psychological framework assumes that this must be explained in terms of a projection of a mental state into that entity. A specific mood might have its ground (in terms of causation, motivation, and justification) in the experience of an entity, but it has become non-sensical within psychological discourse to attribute *Stimmung* to anything else than a mind.

It is important to stress this psychological common sense, because it stands between us and the original meaning of the term *Stimmung*. We need to unlearn to immediately assume that *Stimmung* must be understood in terms of a mental state if we want to comprehend how this term was used in many fields of knowledge. We should also note that classical phenomenologists like Husserl, Scheler, Stein, or Reinach had close ties to the psychologists of their time, thus making them prone to assume the psychological common sense understanding of *Stimmung*. Even Hermann Schmitz, when developing his conceptualisation of feelings as atmospheres (Schmitz 1969, 98–133), mainly refers to sources from early 20th-century psychology and classical phenomenology, sources that are fully committed to the psychological framework. The psychologisation of our thinking has been so influential that it is almost impossible today to understand the term *Stimmung* as signifying anything else than the mental states commonly referred to as moods.

### Art history and life philosophy

At the turn of the 20th century, however, there were still voices that resisted the psychologisation of the term *Stimmung*. Here, I briefly review two such movements. The first concerned the understanding of aesthetic experience. The art historian Alois Riegl (1929 [1899]) drew on the understanding of *Stimmung* which we have seen in Schiller and Schopenhauer when claiming that it is the task of modern art to enable an awareness of order and harmony in nature. Riegl calls this awareness *Stimmung*. Thus, *Stimmung* here denotes a disposition of the mind in which it can experience the entirety of nature as harmonious order. A similar use of the term Stimmung can be found in Georg Simmel's "Philosophy of landscape" (2007). Simmel wonders about the ontological status of landscapes. What is the ground for demarcating a piece of nature as a landscape? Simmel states that "a landscape arises when a range of natural phenomena spread over the surface of the earth is comprehended by a particular kind of unity" (Simmel 2007, 26). According to Simmel, it is the *Stimmung* (maybe here best translated as atmosphere) of a landscape that provides this unity. By means of a specific *Stimmung*, various components are drawn into one unity, which constitutes a landscape. This *Stimmung* is not outside of the components, but it is more than the sum of its parts. The various components only are what they are because of how a *Stimmung* unifies them into a landscape.

Another trail resisting the psychologisation of *Stimmung* is life philosophy (Nietzsche 1994; Bergson 1911; Dilthey 1991). Here, *Stimmung* is understood as synonymous with vital feelings (*Lebensgefühle*) and closely relating to vital power (*Lebenskraft*). Those powers and feelings of life are understood as sub-conscious forces that function as *principium individuationis* for individuals and collectives. The core idea is that different vital feelings are the subsoil of life which allows to explain the emergence of different world views. Accordingly, changing a world view is considered to only be possible via a transformation of the underlying vital forces (Dilthey 1960).

## Three possible English translations of *Stimmung*

The previous section prepares us to understand how Heidegger could draw on the rich conceptual history of the term *Stimmung* from Kant to Dilthey when developing his unique conceptualisation of *Stimmung* as the ontic instantiation of *Befindlichkeit*. Moreover, it prepares us to take seriously that Heidegger's understanding of *Stimmung* is meant as an objection to the psychological common sense that was already firmly established at his time. But before we proceed to Heidegger, it is useful to prepare this discussion further by briefly reviewing three possible English translations of the term *Stimmung*.

### *Mood*

Within a psychological framework, mood is the obvious translation of *Stimmung*. The term mood is from Germanic origin and a version of it can be found in all Germanic languages. The original meaning is "mind, thought, will" (*Oxford English Dictionary* 2001).[3] Hence, mood is fully located in the semantic field of the mind and does not allow for ascriptions outside of a mental context. Therefore, it is unsuitable to render any of the other connotations of the German *Stimmung*.

In current psychological and philosophical debate, the main issue is how moods are distinguished from emotions. Typically, three criteria are identified that might serve as specific differences (Stephan 2017): the first candidate is duration, with the assumption being that moods are typically of longer duration than emotions; the second candidate is the cause, with the idea being that emotions are caused by specific events, whereas moods lack such a specific cause; this relates to the third candidate, which is intentionality, where the idea is that emotions are about particular objects or events, whereas moods are considered as objectless states. However, whereas there is broad agreement about the general nature of emotions, all candidates for distinguishing moods from emotions are disputed. For the context of this paper, it is important to note that moods are unanimously considered to be affective mental states that rather closely resemble emotions. In today's German, *Stimmung* is commonly used as equivalent of mood, and thus, deprived of its rich history outside the confined reference to the affective mind.

### *Tuning and attunement*

By contrast, the semantic field of tuning shares the musical origin of *Stimmung*, and thus, is the obvious candidate to render this original sense of the term in English. This is the reason why I mostly employ tuning to translate *Stimmung*. In the context of a musical instrument, the verb *to tune* is the direct translation of *stimmen*, meaning to adjust the tone of an instrument to the standard of pitch. However, whereas the German *stimmen* stems from

*Stimme*, signifying the human voice, such a relation to the human body is missing in the etymology of tuning. Most likely, *tune* (both verb and noun) is a variant of *tone* (*Oxford English Dictionary* 2001). Like the German *Ton*, it can be related back to the Latin and Ancient Greek *tonus*. Interestingly, *tonus* only figuratively means sound. The original meaning of the Ancient Greek term *tonus* is tension or stretch. The related Verb is *teinein*, meaning *to stretch*. It is only via a metonymy from the process of stretching or straining a string, that *tonus* also came to denote the tone or pitch of something. The original meaning of *tonus* as stretching and straining can still be found today, for example, in the phrase muscle tone. Similarly, the German word for the tension of a muscle is *Muskeltonus*. However, if one wanted to use another German word to translate *Tonus*, it would be *Spannung*, with related verbs like *anspannen* (to tense up) and *entspannen* (to relax). Thus, tuning has a semantic overlap with *Stimmung* in the domain of music, but otherwise relates to a different semantic field.

Besides its use in music, tuning has mostly technical applications. For instance, radio transmission requires a receiver to tune into a frequency, which led to the idiom that one *tunes in* and *out* of radio or TV channels. In engineering, tuning means to adjust the parameters of a system in order for it to achieve certain objectives. For instance, one might want to tune a combustion engine to enhance power output, or to reduce emissions, or to increase durability, or a combination of these and other goals. In natural science, tuning describes the process in which the parameters of a model are adjusted to better fit the observed data. Around 1800, it would have been reasonable to translate tuning in these contexts with *Stimmung*. Today, however, *Stimmung* is so obviously related to sentient beings that it would sound strange to use it in the context of such technical applications.

In Heidegger scholarship, *attunement* has become the standard translation of *Stimmung*. Given the history of the term tuning, I agree with this being the best choice. Most importantly, I consider it highly misleading to translate Heidegger's *Stimmung* as mood. Such a translation completely subverts that it was Heidegger's specific aim to oppose the psychological understanding of *Stimmung*. When translating Heidegger's *Stimmung* as attunement, however, we should note that whereas *to attune* means bringing into accord or harmony (*Oxford English Dictionary* 2001), Heidegger uses *Stimmung* to refer to the entire spectrum of tuning. According to Heidegger, we are always in *Stimmung*, not just when being in tune, but also in cases of being off-tune (*verstimmt*) or supposedly without a tune (*ungestimmt*); these are all modalities of being tuned.

### Atmosphere

Finally, let me briefly discuss the term atmosphere. Atmosphere is a neologism dating back to the 17th century which was built from the Greek terms *atmos* (vapour) and *sphaira* (sphere). It was originally used in a cosmological context, referring to "the spheroidal gaseous envelope surrounding any of the heavenly bodies," and as a special case of that, "the mass of aeriform fluid surrounding the earth" (*Oxford English Dictionary* 2001). Friedlind Riedel (2019) has shown that beginning in the 18th century—hence around the same time in which *Stimmung* became a broadly used thought pattern—atmosphere did not only refer to celestial bodies, but allowed for much broader applications, at least in German and French. It had the general meaning of "corporeal effluvia," designating any kind of vapour stemming from a body. In this sense, it could be applied to the human body, but also to other bodies, for example, signifying the magnetic field surrounding an electrical body. This shows that while the traditional concepts of *Stimmung* and atmosphere developed around the same time, they signify very different thought patterns, with the first building on the semantic field of tuning, the latter on the semantic field of evaporation.

Thus, based on etymology and history, atmosphere is quite far removed from *Stimmung*. So, whereas tuning is best suited to render the original meaning of *Stimmung* in English, and mood is equivalent to the psychological notion of *Stimmung*, atmosphere does not fit any of these bills. On the other hand, atmosphere has been used in recent discourse to describe phenomena that might also be accounted for in terms of *Stimmung*. The current rise of research interest around the term atmosphere is mostly driven by English translations of Gernot Böhme (2017) and the work of Tonino Griffero (2014; 2017; and his contribution in this volume). Both build on but also crucially dissent from the work of Herrmann Schmitz (1969), who introduced an understanding of feelings as atmospheres. According to Schmitz, feelings should not be understood as bound to individuals' minds, but as spatially extended atmospheres that affect individual bodies (Landweer 2020). As I mentioned earlier, it strikes me that Schmitz does not make use of Heidegger's conceptualisation of *Stimmung*, but mostly refers to other classic phenomenologists and early 20th-century psychologist. Moreover, Schmitz, Böhme, and Griffero, at least to my knowledge, do not make use of the conceptual history of *Stimmung* before and beyond the psychological framework. Hence, there is unexplored potential in linking current research on atmospheres with old work on *Stimmung*, as both drive towards subverting binary distinctions between mental and bodily, subjective and objective, inner and outer.

### Heidegger's Befindlichkeit

The previous sections allow us to contextualise Heidegger's account of *Stimmung*. More specifically, the earlier sections enable us to notice how Heidegger could build on previous usages of the term *Stimmung*, especially the understandings encountered in traditional aesthetics and life philosophy. Moreover, it helps us to emphasise that Heidegger's main aim is to subvert the psychological framework in his conceptualisation of *Stimmung*. In addition, the prevision sections contextualise Heidegger's envisioned understand of *Stimmung* within current debates centring around possible English translations. Moreover, it explains the choice of translating Heidegger's *Stimmung* as attunement. In this section, I discuss Heidegger's account of *Befindlichkeit* and *Stimmung*.[4]

### *Befindlichkeit and Stimmung: the basics*

Let me begin by providing a gist of Heidegger's account of *Befindlichkeit*. Following Haugeland (2013), I prefer translating *Befindlichkeit* as *findingness*, but there are other viable options like *disposedness*, or looser but nevertheless useful translations like *situatedness* or *affectivity*. Findingness is one of the four characteristics through which Heidegger in Chapter 5 of Division 1 of *Being and Time* describes the specific way in which Dasein is in the world. The others are understanding (*Verstehen*), discourse (*Rede*), and falling (*Verfallen*). Heidegger emphasises that these characteristics are equiprimordial, that is, they cannot be reduced to each other and they are what they are only in their interplay. Thus, we would need to explore how they together characterise the way in which Dasein discloses a world. In this chapter, I will not explicitly talk about understanding, discourse, and falling, but I will contextualise Heidegger's account of findingness within his overall project.

When Heidegger describes Dasein as being-in-the-world, he makes clear that Dasein's being-in cannot be understood along the same lines as water is in a jar or as a fish is in the water. The world is not a container in which we find Dasein among other things. Rather, Dasein's being-in-the-world means that it is always already engaged with the world, that it always already finds its way around the world one way or the other. Chapter 5 of *Being and Time* is meant to further elaborate what it means to be *in* the world in this specific sense. In this context, Heidegger uses the term findingness to refer to the fact that we always already experience a meaningful world within which things are relevant to us. As soon and as long as we exist, we cannot avoid encountering entities as meaningful and mattering to us. Borrowing a phrase from John Haugeland, we are beings who just cannot stop giving a damn (Adams and Browning 2016). Findingness is Heidegger's term for the facticity of relevance, for the fact that our existence implies a meaningful world within which things matter to us, in which we encounter them as relevant.

Thus, findingness denotes a fundamental ontological structure of Dasein. By contrast, *Stimmungen* (attunements) signify the ontic instantiations of this ontological structure. In other words, whereas findingness refers to the fact *that* things matter to us, attunements allow us to account for *what* matters to us and *how* it matters to us. Our being-in-the-world is always tuned one way or the other. If we want to formulate this idea in a way that is closer to psychological parlance, we can say that attunements disclose at once the relevance of a situation and how it is for me to be in that situation. Attunements serve a world- and self-disclosing function (Slaby and Stephan 2008). Heidegger uses attunement as a generic term to cover a broad spectrum of such world- and self-disclosing dynamics. However, it might have been helpful had Heidegger distinguished different layers of attunement.

First, what Heidegger describes as fear in §30 of *Being and Time* seems to refer to phenomena that are commonly understood as emotions. It does not seem difficult to translate Heidegger's description into today's philosophical parlance: first, fear is based on the encounter with a specific entity (in today's terminology: an emotion's target). Second, fear discloses that entity as fearsome (an emotion's formal object). At the same time, it discloses me or something that is important to me as threatened (an emotion's focus). Finally, fear is of rather short duration as it depends on the felt presence of something fearsome and quickly vanishes once its source is gone. However, in contrast to the standard understanding of emotion shared by most emotion researchers today, Heidegger's terminology of attunement allows him to emphasise that an emotion like fear is only possible because being-in-the-world can be tuned in such a way that it is enabled to discover entities as fearsome or in danger. To play a bit more with the word tuning: fear is only possible for an entity that is tuneable so that it can experience specific entities as threatened and/or threatening. It is important to note, however, that attunement does not only refer to Dasein being affectable in general, but that Dasein's affectability is always modulated in a specific way.

Second, the "undisturbed equanimity" or the "inhibited discontent" (Heidegger 1953, 134),[5] which Heidegger mentions at the beginning of §29, seems to form a second layer of attunement. This layer seems to refer to phenomena that are usually described as moods. Here, it is not a specific entity in a particular situation that is disclosed as such and such. Rather it is a short duration of time that is tuned in such a way that all our comportment is coloured in a specific way, making certain experiences more likely, and certain actions more feasible than others. For instance, in distress being-in-the-world is tuned in a way that makes everything seem difficult, burdensome, and hopeless; activities, that in a different tuning would be easily handled, seem unachievable; situation that a different tuning would reveal to be joyful appear unbearable.

Hence, the first two layers reconstructed from Heidegger's examples of attunements appear to be easily relatable to psychological common sense. However, those are not the kind of attunements Heidegger is interested in. Heidegger's main concerns are those attunements that affect being-in-the-world in a more profound and encompassing way. In *Being and Time*, Heidegger only mentions *Angst* as such an attunement.[6] In other writings, however, it becomes clear that Heidegger considers *Angst* as only one of several such attunements (Heidegger 1998; 1999). What are those attunements and how do they tune being-in-the-world in the most profound way?

### The example of boredom

A good starting point for exploring this issue further is Heidegger's remarks on boredom in the lecture course *The Fundamental Concepts of Metaphysics* (Heidegger 1995). Heidegger distinguishes three forms of boredom, which align with my attempt to relate Heidegger's examples of attunements to a psychological understanding of emotions and moods. In the first form of boredom, one is bored by a specific entity or event. Heidegger uses the example of missing a connecting train and having to wait at a shabby train station. Here boredom discloses something about how it is for an individual to be in a specific situation. An individual is bored about being stuck at a boring train station.

The second form of boredom is more complex. Heidegger's example is an individual who attends a dinner party and appears to participate in a lively fashion. On a surface level, that individual might even experience her own participation in the party as entertaining. But at the same time, the individual is haunted by the intrusive feeling that the dinner party is a dumb activity full of void conversations and that she was bored throughout. In this case, boredom discloses a situation, which on the surface level might seem like a joyful pastime, as deprived of meaning. One might come to ask: What is the sense of dinner parties anyway? Could there be a better way to spend my time?

Finally, in the last form of boredom, one experiences the entire world as deprived of meaningful activity. The world as a whole becomes indifferent; there is just nothing in the world that seems worth pursuing. If one wanted to understand what Heidegger describes here within a psychological framework, one would likely need to come to the conclusion that we are dealing with a case of clinical depression. However, Heidegger maintains that this deepest form of boredom bears profound ontological insight. How is that the case?

## The methodological role of basic attunements

Heidegger takes boredom, like *Angst*, to serve a decisive methodological function. To understand how this is the case, we need to understand Heidegger's overall project. In short, Heidegger aims at understanding how and why we come to raise ontological issue. This meta-reflection on how ontology is possible is what Heidegger calls *fundamental ontology* (Heidegger 1953, 13). *Dasein* is Heidegger's name for the kind of entity that is capable of doing ontology, that is, ask ontological questions and conceiving answers to them.

In Division 1 of *Being and Time* Heidegger shows that Dasein usually takes the world for granted. Everyday Dasein just does not care about ontological issues, because there is no reason to be troubled about the basic structure of the world. However, Heidegger uncovers that Dasein's mundane comportment is made possible by a specific attunement, which might be described as the attunement of familiarity and being-at-home. This attunement discloses the world as a familiar place where one fits in and easily finds one's way around. One is acquainted with the texture of the world in such a way that it does not present major obstacles hindering one's beknown ways of comportment. The world does not provide major puzzles, so there is no reason to be troubled by ontological issues.

How, then, can we be bothered by issues pertaining to the ontological structure of the world? Heidegger claims that we become perplexed about our being-in-the-world if we are seized by attunements that push us out of sync. Heidegger introduces *Angst* and boredom as examples for such attunements that have the power to break our being-at-home in the world, and thus, force us to stop taking the world for granted. The deepest form of boredom achieves this by rendering the entire world deprived of meaningful activities. There is just nothing in the world that matters enough to make engaging with it worthwhile. This might lead one to ask: how is it that meaning comes about in the first place? And suddenly, one is puzzled and perplexed about the fact that we usually are situated in a meaningful world. Similarly, but maybe even more profoundly, *Angst* (as Heidegger understands it) confronts us with the possibility that there could be no meaningful world at all: Why is there a world we encounter as meaningful? Why does anything matter at all? Wouldn't it be just as reasonable to assume that nothing really matters? Based on these examples, we understand Heidegger's claim that we can never be free of attunement and that an attunement can only be overcome by another attunement (Heidegger 1953, 136). This claim is exemplified by how our being-at-home in the world which makes possible everyday comportment becomes interrupted by a retuning (*Umstimmung*) that puts us out of sync.

### Stimmung as world-disclosure

The previous section provided an idea of how attunements are at the heart of Dasein's world disclosure. In and through being tuned, Dasein finds itself situated within a meaningful world. In line with the thought pattern which we have identified in the original musical context of *Stimmung*, we might say that attunements tune the world and Dasein in their reciprocal dependence. Attunements tune world and Dasein in such a way that makes meaningful comportment possible. In this sense, Heidegger can say that we need to "leave the primary discovery of the world" to attunements (Heidegger 1953, 138). This also makes comprehensible that attunements are neither subjective nor objective, neither inner nor outer. Rather, they tune being-in-the-world as a whole.

Against this background, we can appreciate that the ontological depth of Heidegger's analysis does not lie in any "deep layers" of existence, but in the ability to be perplexed about our mundane existence. The world-disclosing function of attunements is related to our everyday comportment in all its modalities. At the same time, world-disclosure cannot be reduced to individual capacities within our mundane comportment. What Heidegger describes as world-disclosing attunements is not a conceptual capacity; one does not need to have any propositional attitude to disclose the world in a specific way. World-disclosing attunements are also not reducible to identifiable emotions or moods. The world-disclosure of attunements cannot be reduced to individual mental states. Rather it denotes the tuning of Dasein and world that makes individual mental states possible.

However, attunements do not signify an openness devoid of specific content. Being-in-the-world is always tuned in a specific way. Our attunements confront us with our situatedness in a specific surrounding and against the background of a concrete past. In and through attunements, we come to encounter the world here and now, in that specific situation with all its historical weight (Slaby 2017). Basic attunements fundamentally shape our being-in-the world. Their gravity goes beyond what can be registered in conscious experience. At the same time, they manifest in the socio-material shape of the world and our most mundane modes of comportment. For instance, Schuetze and von Maur (2021) explored how a rationalistic attunement fundamentally shapes the way in which modern Western societies go about things, tuning all socio-material and affective dynamics within those societies. In a nutshell, the rationalistic attunement shapes our being-in-the-world in such a way that what matters is reduced to what can be quantified. In other words, the rationalistic attunement drives the assumption that every issue needs to be settled by reference to quantifiable parameters. Think about how language proficiency is measured by test scores or how the achievements of scientists are evaluated by publication metrics. These are small examples of how the entire world is tuned towards rationalistic disclosure.

This glimpse at the world-disclosing function of attunements shows that Heidegger's understanding of attunements goes far beyond usual conceptualisations of emotions and modes. Attunements are inscribed into the basic texture of the world, shaping what can come to matter and how it can come to matter.

## Conclusion

A closer look into the conceptual history of the term *Stimmung* reveals that this term was originally used as a universally applicable thought pattern enabling analyses of the (self-)tuning of all kinds of entities in various fields of knowledge. By contrast, the semantic field of *Stimmung* today has become identical to that of mood. Through this semantic shift, the notion of *Stimmung* has not only been enclosed in a narrow semantic field but has also become ontologically trivialised. Heidegger might have been the last one who employed the term *Stimmung* in an anti-mentalistic fashion.

Heidegger's *Stimmungen* go far beyond what can be captured in (affective) mental states. He uses this term to account for how our radical situatedness in the world forms the subsoil against the background of which all our conscious engagements take shape. Understood in such a way, *Stimmung* pertains to a radically dynamic and relational understanding of ontology. If we were to follow this alternative concept of *Stimmung*, it would suggest analysing how the world is dynamically shaped, not just abstractly, but always in a specific way. Following this guideline urges us to investigate the tuning of the world on all levels, from the micro level of particular entities and their interactions, via the meso level of socio-material practices and institutions to large-scale socio-cultural dynamics.

## Notes

1 I speak of conceptual history in a rather colloquial sense and do not refer to the methodology of *Begriffsgeschichte* (Koselleck 2002).
2 I last conducted the search on April 6, 2021. I used the corpus "German (2019)" and the years "1500–2019," searching for the words "Stimmung," "stimmen," and "Stimme."
3 The German version of the term is spelled *Mut* today and means courage. A derivation of *Mut* is *Gemüt*, formed by adding the prefix *Ge-* which indicates a collection or gathering. Thus, *Gemüt* means the unity or entirety of one's mental situation. The term *Gemüt* was frequently used by Kant and his successors.
4 This section builds on my previous work on Heidegger's account of Befindlichkeit (Thonhauser 2017; Slaby and Thonhauser 2019; Thonhauser 2020a; 2020c).
5 As is common practice in Heidegger scholarship, I refer to the page numbers of the German edition of *Being and Time* published by Niemeyer. These page numbers can be found in all English translations of *Being and Time* as well as in volume 2 of the *Gesamtausgabe*.
6 What Heidegger describes as *Angst* is different from what is usually discussed as anxiety in emotion research. For that reason, I prefer to leave *Angst* untranslated to emphasise its difference from the psychological understanding of anxiety.

## References

Adams, Zed, and Jacob Browning, eds. 2016. *Giving a Damn. Essays in Dialogue with John Haugeland*. Cambridge: MIT Press.
Bergson, Henri. 1911. *Creative Evolution*. New York: Henry Holt and Company.
Böhme, Gernot. 2017. *Atmospheric Architectures: The Aesthetics of Felt Space*. London: Bloomsbury.
Dilthey, Wilhelm. 1960. *Weltanschauungslehre: Abhandlungen zur Philosophie der Philosophie. Gesammelte Schriften, Vol. 8*. Göttingen: Vandenhoeck & Ruprecht.
———. 1991. "Introduction to the Human Sciences." In *Selected Works 1*, edited by Rudolf A. Makkreel. Princeton: Princeton University Press.
Geiger, Moritz. 1911. "Über das Wesen und die Bedeutung der Einfühlung." In *Bericht über den IV. Psychologenkongreß in Innsbruck vom 19. bis 22. April 1910*, edited by F. Schuhmann, 29–73. Leipzig: Barth.

Gregg, Melissa, and Gregory J. Seigworth. 2010. *The Affect Theory Reader*. Durham: Duke University Press.

Griffero, Tonino. 2014. *Atmospheres: Aesthetics of Emotional Spaces*. New York: Routledge.

———. 2017. *Quasi-Things: The Paradigm of Atmospheres*. New York: State University of New York Press.

Grimm, Jacob, and Wilhelm Grimm. 1998. "Deutsches Wörterbuch". Digitized version: http://woerterbuchnetz.de/cgi-bin/WBNetz/wbgui_py?sigle=DWB.

Haugeland, John. 2013. *Dasein Disclosed: John Haugeland's Heidegger*, edited by Joseph Rouse. Cambridge: Harvard University Press.

Heidegger, Martin. 1953. *Sein und Zeit*. 7th ed. Tübingen: Niemeyer.

———. 1995. *The Fundamental Concepts of Metaphysics. World, Finitute, Solitute*. Bloomington & Indianapolis: Indiana UP.

———. 1998. "What Is Metaphysics?" In *Pathmarks*, 82–96. Cambridge: Cambridge UP.

———. 1999. *Contributions to Philosophy (From Enowning)*. Bloomington & Indianapolis: Indiana University Press.

Kant, Immanuel. 2007. *Critique of Judgment*. Oxford: Oxford University Press.

Kluge, Friedrich. 2012. *Etymologisches Wörterbuch der deutschen Sprache*. 25th ed. Berlin: De Gruyter.

Koselleck, Reinhart. 2002. *The Practice of Conceptual History: Timing History, Spacing Concepts*. Stanford: Stanford University Press.

Landweer, Hilge. 2020. "Zur Räumlichkeit der Gefühle." *Gestalt Theory* 42 (2): 1–15.

Lipps, Theodor. 1903. *Ästhetik*. Leipzig: Voss.

———. 1909. *Leitfaden der Psychologie*. 3rd ed. Leipzig: Engelmann.

Lotze, Hermann. 1868. *Geschichte der Ästhetik in Deutschland*. München: Cotta.

Nietzsche, Friedrich. 1994. "Über Stimmung." In *Frühe Schriften. Band 2*. München: Beck.

*Oxford English Dictionary*. 2001. Oxford: Oxford University Press. http://www.oed.com/.

Prandtl, Antonin. 1910. *Die Einfühlung*. Leipzig: Barth.

Riedel, Friedlind. 2019. "Atmosphere." In *Affective Societies: Key Concepts*, edited by Jan Slaby and Christian von Scheve, 85–95. New York: Routledge.

Riegl, Alois. 1929. "Die Stimmung als Inhalt der Modernen Kunst." In *Gesammelte Aufsätze*, 27–37. Augsburg und Wien: WUV.

Schiller, Friedrich. 1982. *On the Aesthetic Education of Man in a Series of Letters*. Oxford: Clarendon Press.

Schmitz, Hermann. 1969. *System der Philosophie, Teil III.2. (Der Gefühlsraum)*. Bonn: Bouvier.

Schopenhauer, Arthur. 1972. *Die Welt als Wille und Vorstellung. (Sämtliche Werke 2)*. Wiesbaden: Brockhaus.

Schuetze, Paul, and Imke von Maur. 2021. "Uncovering today's rationalistic attunement." *Phenomenology and the Cognitive Sciences*. https://doi.org/10.1007/s11097-021-09728-z.

Simmel, Georg. 2007. "The Philosophy of Landscape." *Theory, Culture & Society* 24 (7–8): 20–29.

Slaby, Jan. 2017. "More than a Feeling: Affect as Radical Situatedness." *Midwest Studies in Philosophy* 41 (1): 7–26.

Slaby, Jan, and Achim Stephan. 2008. "Affective Intentionality and Self-Consciousness." *Consciousness and Cognition* 17 (2): 506–13.

Slaby, Jan, and Gerhard Thonhauser. 2019. "Heidegger and the Affective (Un) Grounding of Politics." In *Philosophers in Depth: Heidegger on Affect*, edited by Christos Hadjioannou, 265–289. London: Palgrave Macmillan.

Stephan, Achim. 2017. "Moods in Layers." *Philosophia* 45: 1481–1495.

Thonhauser, Gerhard. 2017. "Transforming the World: A Butlerian Reading of Heidegger on Social Change?" In *From Conventionalism to Social Authenticity: Heidegger's Anyone and Contemporary Social Theory*, edited by Hans Bernhard Schmid and Gerhard Thonhauser, 241–260. Cham: Springer.

———. 2020a. "Authenticity and Critique: Remarks on Heidegger and Social Theory." In *Regelfolgen, Regelschaffen, Regeländern—Die Herausforderung für Auto-Nomie und Universalismus durch Ludwig Wittgenstein, Martin Heidegger und Carl Schmitt*, edited by Matthias Kaufmann and James Thompson, 115–131. Bern: Peter Lang.

———. 2020b. "Beyond Mood and Atmosphere: A Conceptual History of the Term Stimmung." *Philosophia* 49:1247–1265. https://doi.org/10.1007/s11406-020-00290-7.

———. 2020c. "Martin Heidegger and Otto Friedrich Bollnow." In *The Routledge Handbook of Phenomenology of Emotions*, edited by Thomas Szanto and Hilge Landweer, 104–14. London & New York: Routledge.

Volkelt, Johannes. 1905. *System der Ästhetik*. München: Beck.

Wellbery, David E. 2003. "Stimmung." In *Ästhetische Grundbegriffe (ÄGB): Historisches Wörterbuch in Sieben Bänden*, 7: 703–733. Stuttgart: Metzler.

Welsh, Caroline. 2008. "Nerven—Saiten—Stimmung. Zur Karriere einer Denkfigur zwischen Musik und Wissenschaft 1750-1850." *Berichte zur Wissenschaftsgeschichte. Organ der Gesellschaft für Wissenschaftsgeschichte* 31 (2): 113–129.

———. 2009a. "Die ‚Stimmung' im Spannungsfeld zwischen Natur- und Geisteswissenschaften. Ein Blick auf deren Trennungsgeschichte aus der Perspektive einer Denkfigur." *NTM. Zeitschrift für Geschichte der Wissenschaften, Technik und Medizin* 17: 135–69.

———. 2009b. "Resonanz—Mitleid—Stimmung: Grenzen und Transformationen des Resonanzmodells im 18. Jahrhundert." In *Resonanz. Potentiale einer akustischen Figur*, edited by Karsten Lichau, Rebecca Wolf, and Viktoria Tkaczyk, 103–22. München: Fink.

Wiener, Norbert. 1948. *Cybernetics: Or Control and Communication in the Animal and the Machine*. Cambridge: MIT Press.

Witasek, Stephan. 1904. *Grundzüge der allgemeinen Ästhetik*. Leipzig: Barth.

# 3 Moods and atmospheres

## Affective states, affective properties, and the similarity explanation

*Íngrid Vendrell Ferran*

## Introduction

In ordinary language, "calmness," "melancholy," "cheerfulness," and "sadness" are employed to describe affective states experienced by sentient beings. More precisely, these terms are used to report instances of moods. Yet, the very same terms are used to describe what seem to be properties of certain objects (e.g., things, situations) which, unlike sentient beings, are unable to feel. We usually describe atmospheres employing these terms. For example, we speak about the calmness of a forest, the melancholy of a painting, the cheerfulness of a field of flowers, and the sadness of a landscape.[1] This double connotation raises the following set of questions: Are we, in fact, referring to the same phenomenon? If not, why then do we employ the same terms?

In order to answer both questions, I proceed as follows. In the first section, I offer an accurate analysis of the structure of moods and of atmospheres, arguing that both phenomena are distinct in kind. Moods are affective states, atmospheres are affective properties. This "distinctiveness thesis," as I call it, provides an answer to the first question posed above. Next, I focus on how atmospheres as affective properties are apprehended and experienced. In particular, I defend the "model of feeling" according to which atmospheres are apprehended by an intentional feeling. I proceed then to examine the wide array of experiences which the feeling of atmospheres might elicit. In the last part of the chapter, I provide an answer to the second question previously mentioned. In particular, I argue that terms for atmospheres are borrowed from the vocabulary of moods on the basis of a similarity between both phenomena. I call it "the similarity explanation." The main findings are summarised in the conclusion.

## Distinguishing moods from atmospheres: the distinctiveness thesis

Our experience of moods and atmospheres differs substantially. When we claim to feel sad, we refer to our own affective state. By contrast, when we speak of the sadness of the landscape, we describe a phenomenally objective

DOI: 10.4324/9781003131298-4

property of the landscape, i.e., a property which is experienced as being objectively there. In what follows, I will take this experiential difference at face value and examine essential traits inherent to the structure of each of these phenomena. The aim is to pinpoint those features that enable a comparison between both phenomena.

## Moods as affective states

I begin by fleshing out the structure of moods as a kind of *affective state*. In particular, I am interested here in their distinctive intentional structure.[2] Using the emotions as a foil for comparison, I will distinguish three different kinds of objects of moods (for comparisons involving other elements, see Krebs 2017 and Rossi 2019).[3]

In emotion research, it is customary to distinguish between material and formal objects of the emotions (Kenny 1963, 193). Material objects are the targets of the emotions (broadly speaking: things, animals, persons, situations, etc.) which as such can be quite idiosyncratic, culturally and socially learned, and individually variable. By contrast, the formal object refers to the evaluative dimension in which these targets are presented to us. For instance, fear can be directed towards a thing, a person, an animal, a situation, etc., but its formal object is always the same: regardless of the targeted objects, they are all presented as dangerous.

Broadly speaking, a parallel case can be drawn for moods. Just as in fear (emotion) an object is dangerous, in sadness (mood) things seem less enjoyable and bathed in black. Like emotions, moods can also be regarded as being directed towards something and presented in a certain evaluative light. Yet, this parallel is obviously imperfect and includes some intriguing differences.

First, unlike emotions, the material objects of moods lack specificity. While my fear targets a certain object, my sadness affects everything. There have been different accounts of moods' lack of specificity; for instance, it has been claimed that moods target all that I encounter while being in this mood (Sizer 2000), the world (Crane 1998) or everything and nothing in particular (Goldie 2000, 8 and 148; Solomon 1993, 71). All these different accounts point to the fact that unlike emotions, moods do not have specific objects.

Second, though both—emotions and moods—have an evaluative dimension, their evaluative character differs considerably. Emotions respond to objects that are presented as having certain evaluative properties,[4] while moods do not respond to anything in particular; rather, moods illuminate with their light the different objects that they might target. Moreover, emotions are connected to specific evaluative properties (fear responds to the dangerous), while moods are related to different kinds of evaluative properties. When I am sad, I do not apprehend all the objects that I can target while being in this mood as having a unique evaluative property. What happens instead is that in sadness, we tend to be more sensitive to certain clusters of evaluative properties such as the menacing, the depressing, etc. than others (my understanding of moods here is close to the understanding of moods in terms of likelihoods, [see Price 2006, 51]).

The distinction between two kinds of objects is crucial to understand the structure of the emotions and, in my view, the structure of moods, too. However, the intentional structure of emotions and moods is not exhausted by this distinction. Emotions and moods do not leave their objects unaffected but make them appear as having a certain aspect according to the emotions or mood we are experiencing. They have the ability to impress their characteristic lights on their targets. As we put it in common parlance, affective states "colour," "tincture," and "dye" their targets. Geiger, who observed this feature, argued that affective states add a colouration (Färbung) to the targeted objects, giving them a brilliance (Glanz). He calls it a "feeling tone" (Gefühlston) which is to be understood as spreading over the objects targeted by the affective states (Geiger 1976, 26).[5] My thought here is that this colouration or light can be considered as a third kind of object of emotions and moods.[6] Each emotion and each mood has its specific characteristic colouration which spreads over the targeted objects. Thus, in fear (emotion), an object is presented not only as dangerous but also as having a certain colouration spreading over the targeted object and its surroundings. When I am sad (mood), this sadness tinctures all that we come across with a specific light. However, regarding this third object, emotions and moods differ in their scope. As a result of targeting everything we encounter while being in a mood, when it comes to spreading their colourations, moods have a wider reach than emotions whose objects are more circumscribed.

From these considerations, the following picture of the structure of moods emerges. Moods target everything we encounter while being in a mood (material object); they make us more prone to experience certain sets of evaluative properties (formal object); and they have the ability to impress their specific colouration onto the world (third object). To give an example: when I am sad, this sadness targets everything I encounter while being in this mood; it makes me more sensitive to grasping clusters of negative properties such as the dangerous, the menacing, etc.; and it casts its specific dark colouration over its objects.

Nothing similar to this structure can be found in the phenomenon of atmospheres. When we claim that "the landscape is sad," the sadness that we find "in" the landscape does not exhibit the typical structure of moods presented above. Though the sadness is experienced as belonging to the landscape, the sadness does not target the landscape (it does not have a material object), it does not make us sensitive to certain evaluative properties (it lacks formal objects), and it does not colour anything beyond itself. In fact, it seems that the atmosphere by itself has a colouration and that this colouration spreads over the landscape. However, this colouration is not part of an intentional structure, as was the case for moods as affective states. In short, the sadness of the landscape does not exhibit the distinctive intentional structure of moods as affective states.

## Atmospheres as affective properties

From the previous considerations, it is clear that when we speak about "the sadness of the landscape," we are not describing an affective state. Rather, we are describing the way in which an object, in this case the landscape, appears to us. This appearance is what we often call aura, character, and also atmosphere (which is the terminology that I adopt in this chapter). How to determine the atmosphere's structure?

To begin, some phenomenologists and phenomenological inspired authors attribute to atmospheres a certain degree of reality as "half-things" (Böhme 2001, 45 and 2006, 19; Griffero 2014 and 2017, both taking the idea from Schmitz 2005, 343 and 2008, 269).[7] Although I think that atmospheres have a degree of reality, I will not follow this line of thought here. Instead, I will start by considering atmospheres in terms of *properties*. To speak about "the sadness of the landscape" is to speak about a property of the landscape. This property is experienced as being "in" the object, or, more precisely, as "spreading over" the object or as "covering" it. We neither "interpret" the landscape as sad, nor does the landscape "express" sadness or "evoke" sadness in us. The "sadness" is literally to be found in the landscape as one property of it.

First, an important feature of atmospheres is that they are experienced as issuing from the object rather than as arising from the self. Atmospheres are experienced as having phenomenal objectivity, as being objectively there and as such as accessible to others. People usually converge about the existence of an atmosphere. This experience is what has been called "emotional realism" (Osborne 1968, 105; see also Hepburn 1968, 82; Morris-Jones 1968, 99).

Second, atmospheres are properties of a certain kind. They are what I call here "affective properties." (Though I will use both concepts—atmosphere and affective property interchangeably, not all affective properties are atmospheres. For instance, if we claim that the "landscape is impressive," we are speaking here also about an affective property of the landscape. However, "impressive" unlike "sad" is not a term borrowed from our vocabulary of moods. Rather, it is borrowed from the vocabulary for general feelings.) These properties are affective because they are experienced as related to our affective life. They are experienced as having the capacity to move and affect us, to change our position within the world and to orient us accordingly. Atmospheres make the world a place with reliefs, nuances, surfaces and profundities, colour and light.

Third, atmospheres as affective properties are dependent and founded on other properties of the object. For an object, e.g., a landscape to look sad, it must fulfil certain conditions such as having a particular form, colour, etc. If we change these other features of the landscape, then the look, appearance, of the landscape will change too.[8] By virtue of this feature, atmospheres can be designed and created. This happens by modifying and preparing the non-affective properties of the objects on which they are founded. Moreover, their temporal structure, existence and specific aspect or colour will depend on these other properties of the object.

Finally, affective properties have an evaluative character. Like axiological properties such as beautiful or unfair, affective properties call us to adopt a pro- or contra-attitude towards the object that possesses them.

What this analysis of the structure of atmospheres reveals is first of all that their structure differs substantially from that of moods. This supports "the distinctive thesis" according to which both phenomena are distinct in kind. More specifically, the analysis shows that moods are affective states with a distinctive intentional structure directed towards three kinds of objects, while atmospheres are affective properties founded on other non-affective properties of the objects in relation to which they are experienced. We can now answer the first question raised in the introduction: although we employ the same names for both phenomena, moods and atmospheres can be sharply distinguished. To answer the second question requires a further analysis of atmospheres. In particular, the question about the apprehension and experience of atmospheres has to be carefully analysed.

## How are atmospheres apprehended? The model of feeling

The understanding of atmospheres as affective properties defended above raises an urgent epistemic question: How are atmospheres apprehended? After scrutinising the models of perception, recognition, emotion/mood, and empathy,[9] I will argue that atmospheres are apprehended by a feeling.

### *Perception*

We often employ the language of perception for the apprehension of atmospheres. We speak of seeing the sadness of the landscape or the melancholy of a painting. The perception model takes this manner of speaking at face value and argues that atmospheres are perceived. The perception involved here is not plain sensory perception (e.g., the perception of trees and chairs), but a *sui generis* perception, one which is able to grasp the affective nuances of an object.

One prominent exemplar of this model is Wollheim's account of "expressive perception" as perception of expressive properties (another name for what I call "affective properties"). In his view, an expressive perception is a genuine form of "seeing" just as "seeing-in" (e.g., seeing Napoleon in a painting) is also a sui generis form of seeing appropriate to representation. However, while "seeing-in" enables us to see something represented in a picture, expressive perception makes it possible to see the picture's expressive qualities. As conceived by Wollheim, expressive perception presupposes beliefs derived from certain experiences of the world (this feature is common to other forms of seeing) and it presupposes a mechanism for coping with feelings, moods, and emotions (1987, 80). For him, expressive properties are not entirely independent of our affectivity. In fact, they are the result of a complex projection of our feelings onto the world (see below).

One of the strengths of the perception model is that it captures not only the way in which we usually speak about atmospheres and affective properties, but also the immediacy and directedness in which this apprehension takes place. However, my main concern with this model is that perception is not an intrinsically affective phenomenon, while the apprehension of affective properties is itself affective. In affective states, we are affected and moved in a way in which reality is presented under a certain evaluative light.[10] This involves a twofold moment: the experience of reality in evaluative terms and the experience of a change in our position in the world. Both moments appear in all phenomena belonging to the affective family.[11] My thought here is that the apprehension of the sadness of the landscape involves these two aspects: while experiencing the sadness of the landscape as an evaluative feature of it, I also experience a change in my relation to the world. It is precisely this affective character of the apprehension of affective properties that the perception model cannot capture. Although Wollheim acknowledges that this perception involves, evokes, and projects affective states, perception is not a phenomenon that belongs to the family of the affective (constituted by emotions, moods, feelings, etc.). In my view, we should find among the different phenomena belonging to the affective family a suitable candidate that explains the apprehension of affective properties.

## *Recognition*

According to this model, the apprehension of atmospheres in particular and of affective properties, in general, can be explained as a form of "recognition." In this model, to see the sadness of the landscape is in fact to recognise sadness in it. The recognition involved does not involve judgement, nor should it be taken in a literal sense as a re-cognition (i.e., as identification of something already familiar to us). Recognition is here an activity of the mind akin to understanding.

This model was prominently defended in aesthetics by Hepburn and Morris-Jones. For Hepburn, there is a reasonable extension of "seeing" affective properties (in his terminology: emotional qualities) which is recognising them. Though we can recognise an affective property without being ourselves in the same affective state, he does consider the occurrence of highly particularised feelings involved in this recognition (1968, 91). For Morris-Jones, the recognition requires neither that we experience feelings similar to the apprehended affective properties, nor that we are already acquainted with them. In his view, the conditions for apprehending such feelings are similar to those for the understanding of statements made in a language one has learnt. For Morris-Jones, one learns to talk not only by learning the grammatical rules but also by developing a sensitivity as to when to make certain statements. In a similar vein, we can recognise affective properties we have never experienced (this presupposes that art is a kind of language of feeling whose norms and uses we have learnt): "Recognizing a feeling is like understanding the meaning of a statement, and the conditions of understanding are the same as those which govern the apprehension of feeling" (Morris-Jones 1968, 103).

Rightly, this model points to the ability of the subject to apprehend affective properties and explains this ability in terms of the active capacity similar to understanding. However, the model faces a series of difficulties. First, as was the case with the model of perception, the model of recognition explains the apprehension of atmospheres by resorting to a cognitive phenomenon. Second, a more pressing difficulty is that the model is conceived mainly for art works. The analogy between recognising affective properties in art and understanding language is central to this model. In fact, the model of recognition takes art to be a form of language with its conventions and systems of symbols that others might create and interpret correctly. Yet, affective properties permeate our experience outside and beyond artistic contexts. Finally, those not familiar with the same conventions and symbols as us can also be capable of apprehending affective properties.

### Affective states: emotion/mood

This model explains the apprehension of atmospheres as affective properties in terms of affective states such as emotions and/or moods. For the proponents of this model, the apprehended affective properties are identical to the affective states elicited in the observers. More precisely, emotions and/or moods are responsible for disclosing such properties. In the past, this view was defended by Aldrick and Baensch (see Osborne 1968, 109). Although today this model has not been developed to explain the apprehension of affective properties, perceptual models of emotions (Tappolet 2000) and moods (Rossi 2019) embody a similar idea. By arguing that emotions and/or moods are able to present or represent evaluative properties, these models attribute to affective states the strong epistemic function of grasping sensible aspects of reality, and among the latter affective properties have a place.

Unlike the preceding models, this model regards the apprehension of affective properties as affective phenomena. However, this model also has to face a series of objections. First, to apprehend the sadness of a landscape is not the same as feeling sad ourselves. We are very well able to apprehend the sadness of a landscape without being in the same affective state. It is even possible to apprehend the sadness while being in the opposite affective state. We can apprehend the sadness of the landscape and be aesthetically pleased by it.[12] While it is possible for us to end up sad after apprehending the sadness of the landscape, this sadness is either a case of affective contagion or the causal effect of apprehending the sadness. But in itself it is not "the way" in which we grasp the sadness of the landscape. Thus, affective states cannot be responsible for the apprehension of atmospheres.

Moreover, this model falls prey to a crude oversimplification of reality. For this model, irritability presents offensiveness and anxiety presents threat. This might seem *prima facie* plausible. However, this does not capture what really happens in ordinary experience. When I am irritable, things are not just offensive. They are also threatening, menacing, obnoxious, etc. This seems to be a rule of thumb for all moods we can experience. For instance, euphoria is related to the wonderful, joyful, awesome, pleasant, outstanding, etc. Sadness is related to the threatening, menacing, inhibiting, unpleasant, etc. (This confirms what I have already set out in above; namely, each mood is associated not with one but with many evaluative properties.)

The last two arguments demonstrate that affective states cannot apprehend the affective properties of the objects they are directed towards.[13] A third objection can be derived from the ontology of the mind. Affective states such as emotions and moods have a temporal duration: they extend over time and have a course of development which is compounded by different temporal moments. By contrast, the apprehension of an affective property exhibits the features typical of punctual occurrences.[14] The apprehension of the sadness of the landscape takes place at a given moment and although it can take place repeatedly, it neither extends over time nor has temporal parts. In this regard, the apprehension of an atmosphere has the character of an activity of the mind, something that we achieve as a result of something we do (though not necessarily something we intend). As such it is akin to other activities such as perceiving or recognising (to mention here two of those mentioned above), rather than to conditions of the mind such as emotions and moods.

### Empathy

In this model, atmospheres are apprehended through empathy. This idea was put forward by Lipps for whom the concept "Einfühlung" means literally "feeling into" something.[15] Given that objects, unlike sentient beings, are unable to feel, it is by virtue of feeling into them, i.e., of projecting ourselves into them, that we come to experience in them affective properties.

The most detailed development of this model can be found in Geiger, who developed an account of the empathy of moods (*Stimmungseinfühlung*). In his view, we can "feel into" atmospheres (to which he mainly refers as "feeling characters" (*Gefühlscharakteren*) or simply "characters" (*Charakteren*)). When this happens, we can remain in a contemplative attitude by means of which we experience the atmosphere as being objective and external, or we can adopt an "immersive attitude" in which we immerse ourselves in the atmosphere.[16] Geiger distinguishes four kinds of immersion: 1) in "objective immersion," we open ourselves up to the atmosphere and are able to experience it but we remain passive towards it; 2) in "position-taking immersion," we apprehend the atmosphere and this apprehension, in turn, influences the way in which we perceive it (there is here an interplay between the atmosphere and my own affective state); 3) in "sentimental immersion," we apprehend the atmosphere with the aim of resonating with it: we want to recreate in us the affective state that corresponds to the atmosphere (the atmosphere and affective state converge in a mood experienced by the subject); 4) in "empathic immersion," we can feel into the affective property so that we are completely absorbed by the atmosphere and become one with it.

This model captures crucial features of the apprehension of affective properties. It understands the apprehension of affective properties as affective phenomenon. Moreover, in this model, the apprehension of affective properties has the character of an activity and an achievement. In this respect, the model is superior to those examined above. However, the model conceives of this affective activity as a "feeling into," commonly translated as empathy. The problem here is that we are able to disclose affective properties even if we do not feel into them. Geiger's contemplative attitude confirms this possibility. Following this idea, I will defend the view that affective properties are felt, although this does not necessarily presuppose that we feel into them.

## Feeling

This model is closely related to the previous one. Here affective properties such as atmospheres are apprehended through a feeling. With this term, I refer to an *intentional* affective phenomenon by means of which we are able to apprehend the sensible aspects of reality.

The view is present in Geiger (1911) and in Scheler. Scheler distinguishes the feeling of affective properties (in his terms: mood-characters (*Stimmungscharaktere*)) from the feeling of axiological properties (e.g., dangerous, unfair). In his view, both feelings are intentional but only the feeling of axiological properties (or values) is intentional and cognitive (Scheler 1973, 257).[17] Outside the phenomenological tradition, Osborne developed the idea of a "cognitive feeling"[18] of affective properties (in his terms: emotional qualities). As he puts it: "We *feel* that a work of art is dainty, austere, florid, compact. But the feeling is not awareness of an emotional response in ourselves, deflecting attention inwardly, it is an outward directed, cognitive feeling" (1968, 116).

The model of feeling constitutes, in my view, a plausible alternative to the models discussed above. It captures the crucial features of the apprehension of affective properties stated above. First, the model of feeling captures the affective nature of the apprehension of affective properties. In apprehending an affective property, we are affected and moved. Reality appears not as neutral, but as a relief with certain features in the foreground and others in the background, with fields coloured in one way or another. Things are not neutral, but tinctured with a certain colouration, for instance, as being "sad," "melancholic," or "cheerful." This, in turn, makes us change our position within the world because we orient ourselves according to what we apprehend. Indeed, this intentional feeling exhibits the two features that I considered crucial to determine affective states: we are affected and moved in a way in which reality is presented under a specific "evaluative" light and we experience a change in our relation to the world.

Second, the model of feeling has the advantage that it underscores the active nature of the apprehension of affective properties. Unlike affective states which are conditions of the mind, this intentional feeling is an activity oriented towards an achievement: the grasping of a sensible feature of the world.

Third, this feeling is a punctual occurrence with a minimal temporal duration. It happens in time, but it does not stretch over time, though we might feel the property again and again. More specifically, the feeling of affective properties resembles other punctual occurrences such as "noticing" rather than mental states or processes that occupy stretches of time. Just as we can notice that there is a bird hidden in a tree, we can feel that the situation is sad.

Finally, the model indicates the existence of a form of receptivity towards sensible aspects of reality.[19] This feeling indicates the existence of an ability to sense certain aspects of reality. We are equipped with a sensitivity for evaluative properties such as affective properties (e.g., cheerful, sad) and axiological properties (e.g., unfair, dangerous). The fact that we possess this sensitivity makes us able to orient ourselves in the world in action and thought.

## Experiencing atmospheres: affective stances and affective responses

While in the previous section I have described the apprehension of atmospheres in terms of feeling, this section focuses on the wide range of experiences that this feeling might evoke. In particular, I am interested in describing three affective stances that we might adopt towards our own feeling of atmospheres and in fleshing out some of the most common affective responses that this feeling might elicit.[20]

As noted in classical phenomenology, we are able to adopt an affective stance towards a wide range of our own affective experiences (Geiger 1974 and 1976; Haas 1910). This stance must not be consciously and reflexively adopted. It is also not an evaluation or judgement about what we feel. Rather, it is an affective "yes" or "no" towards our affective experiences. We might be open or closed towards what we experience. For the case that occupies me in this chapter, this means that we can adopt a stance towards the feeling of atmospheres. This stance is what determines whether or not the feeling of atmospheres will unfold and evoke other affective experiences.

Prima facie, three main stances can be distinguished.

### Neutral

It is possible to feel an atmosphere and remain neutral towards this feeling. This happens when we feel the sadness of the landscape, but this feeling does not elicit a response in us. The neutral stance is typical for cases in which we are unaware of the feeling despite registering the affective property. Neutrality might also happen when we are aware of ourselves feeling the atmosphere, but we remain indifferent about it.

### Analytic

A second possibility consists in adopting an analytical attitude towards the feeling of an atmosphere. This attitude aims at examining different facets of the feeling. It can be directed towards some properties of the feeling such as its intensity, depth, quality, potential to elicit various emotional responses, etc. However, we can also focus on the affective property presented by the feeling itself. In this case, the analysis targets what we feel (the content) rather than on how we feel it (the mode). Our analytical stance might target the qualitative dimension of the property, its temporality, the elements upon which it is based and on which it depends, etc. We can attend to the sadness of the landscape and try to scrutinise its specific colouration, the object that it covers, which elements of the landscape it depends on, which elements are essential to it and which are contingent, and so on. A common result of adopting the analytical stance is that the feeling of the atmosphere usually does not unfold, or at least its development and its capacity to elicit emotional responses stagnate momentarily.

### Responsive

Finally, it is also possible to adopt a responsive stance towards the feeling of the atmosphere. Here the feeling unfolds and elicits a wide array of affective responses. These responses might be second-order feelings as well as emotional reactions.[21] In the first case, we have a feeling about the feeling of an atmosphere. For instance, in feeling the sadness of a landscape, we might like or dislike this feeling, we might feel comforted or discomforted by it, etc. In the second case, the feeling of the atmosphere evokes an emotional reaction. Feeling the sadness of the landscape might elicit sadness, melancholy, or even joy (as in the case when the sadness is apprehended within the context of an aesthetic experience).

In all these cases belonging to the responsive attitude, the subject resonates with the atmosphere while at the same time being aware of the difference between one's own affective states elicited by the feeling of atmosphere, and the affective properties which are presented by this feeling. However, in limited cases, the awareness of this difference might disappear. In emotional infection, the subject comes to experience an affective state which is similar to the apprehended affective property without being aware of the reasons why she is feeling this way. The subject might also feel one with the atmosphere and be totally absorbed by it (Schloßberger 2019).

## The similarity explanation

Previously, I traced the distinction between moods and atmospheres in terms of a difference between affective states and affective properties. In the last two sections, I demonstrated that nonetheless affective phenomena play a crucial role in the apprehension and experience of atmospheres. In this final part, I want to tackle the second question posed in the introduction: why do we employ the same terms for moods and atmospheres if they are, in fact, distinct phenomena? It is puzzling that mood terms are employed to describe properties of objects which are unable to feel themselves. This is particularly intriguing because—as I have shown—these properties are not apprehended by homonymous affective states. I argued that they are apprehended by a feeling, but this feeling is not a mood.[22] The landscape does not feel sad, nor do we apprehend this sadness by feeling sadness ourselves. Why then do we speak of the sadness of the landscape?

Possible answers to this question have been divided mainly into two camps (for an overview which includes other explanations, see Krebs 2017, 1423–1427). On the one hand, proponents of "the causal explanation" argue that affective properties are ascribed to an object in accordance with the affective states that the object might elicit or cause in us. The relation of causality can be conceived in either direct or dispositional terms. According to the "direct" causation explanation, that the landscape is sad means that it causes sadness in us. As noted above, this explanation fails because we can apprehend the sadness of the landscape without being sad. In its "dispositional" version, we attribute affective properties to an object when this object predisposes us to experience certain emotional responses. We call a landscape sad when it has the potential to elicit sadness in us. Though this version avoids the main problems of the direct causation explanation, it overlooks the fact that the apprehension of an affective property is not always linked to the experience of similar affective states. As noted above, we might perceive sadness in the landscape, but end up experiencing emotions which differ strongly from sadness (e.g., we might aesthetically enjoy it).

On the other hand, there is the "projectionist explanation" which comes also in two forms. The "simple" version explains affective properties as a projection of our affective states onto an object. Affective states colour and tincture our world. The landscape is sad because I project my sadness onto it. Though, as stated above, affective states have the capacity to colour their targets, the simple explanation is flawed since we can apprehend affective properties without being ourselves in a similar state.

The "complex" version of projectionism was provided by Wollheim. The model of expressive perception mentioned above involves a process of complex projection in which "the emotion flows from what we perceive to us" (Wollheim 1987, 82). Though this emotion is apprehended independently of our present affective state, complex projection involves simple projection. Drawing on Freud, Wollheim describes the mechanism of projection as an unconscious process in which phantasy is involved. There is an initial phantasy motivated by an emotion we want to retain or get rid of. Here the emotion is expelled from the body and spreads across the environment. Furthermore, the subject who expels the emotion has the disposition to fantasise that the world can be experienced in a certain way. In his view, here a "doubling-up of the predicate" takes place (Wollheim 1987, 84). This doubling-up is an example of how language can be idealised and how it might lead us to speak metaphorically about the world so that we end up misunderstanding our experience. This idealisation might lead us to understand our ability to apprehend affective properties as "the metaphorical application of psychological predicates to the world" (Wollheim 1987, 85). Though this account points to the fact that affective states are able to colour the targeted objects, it rests upon a problematic idea according to which we are able to expel affective states which we later experience as having phenomenal objectivity.

I turn now to what I think is the most plausible alternative: the "similarity explanation." The roots of the similarity explanation can be found in Geiger.[23] For this author, each mood projects a specific glow, light or brilliance onto the objects it targets, which he calls "feeling tone," and this glow, light or brilliance is *qualitatively similar* to "feeling characters"—a term he employs to refer to atmospheres (Geiger 1976, 36).[24] In what follows, I will elaborate on this explanation by examining the relation of similarity in more detail.

To begin, the similarity does not concern the mood and the atmosphere in their totality (as we have seen, we can apprehend an atmosphere despite being in a different affective state). Rather, the similarity concerns exclusively certain aspects of each of these phenomena. Put otherwise: certain aspects of moods are similar to certain aspects of atmospheres. In order to determine these aspects, we need to come back to the main features of the structure of moods and atmospheres developed above. Are any aspects of their respective structures similar?

At the level of the material object, no similarity can be stated between moods and atmospheres. Though both phenomena are not punctually localised and spread over several objects, the relation with these objects differs considerably. While affective states have an intentional structure, affective properties are based and depend on other properties of the objects in which we experience them. My sadness might be directed towards everything that I encounter while being in this mood, while the sadness of the landscape does not target any of the objects upon which it is based.

Next, the similarity does not concern the evaluative dimension of each of these phenomena. Moods make us more sensitive to the apprehension of a

certain cluster of evaluative properties, while atmospheres are only evaluative in the sense that they call us to adopt certain affective responses (pro- or
contra-attitudes) towards them.

More promising is the analysis of moods' third object that I called above
the colouration characteristic of each affective state impressed upon the
objects that state targets. In the view that I proposed, each mood has a
typical colouration (the concept of colouration means not just one colour
but a sum of features such as grey, black, wilted, lifeless, etc.). On the other
hand, as described above, in our everyday experience, the world is neither
experienced as a neutral surface nor is it always apprehended in the same
manner. The way in which we experience reality changes. Reality has different aspects and looks. In this chapter, I argued that one good way to
capture these different looks of reality is to speak about affective properties
and atmospheres as having typical colourations. This typical colouration
which is founded on other properties of the object and which is responsible
for presenting the world as a relief according to which we can orient ourselves in action and thought. The only point of convergence between both
phenomena is precisely this one: both involve a colouration. With this term,
I refer to how things look when we are in a mood (moods' third object) and
how atmospheres afford a certain look to things.

Now the similarity explanation runs as follows. Sometimes we might
realise that the world looks similar to how it looks when I am in a certain
mood (e.g., sadness). As a result, and due to the lack of a rich vocabulary
to describe all sensible aspects of reality, we resort to the vocabulary of
moods and name this aspect of the world by employing the same term (e.g.,
sad). We literally borrow the vocabulary of moods to describe atmospheres.
What is similar is this colouration that each mood typically projects onto
the world and the aspect or look characteristic of each affective property.[25]
In other words, the colouration that I project onto the world when I undergo
an affective state is similar to the colouration (aspect or look) of the world at
certain times. To give an example: in sadness, I project onto objects a characteristic colouration which is similar to how these objects look on certain
occasions. It is on the basis of this similarity that I then call these occasions
"sad."

Like Wollheim's explanation, the similarity explanation acknowledges
the ability of affective states to project their typical colouration onto the
world. Both explanations share the view that we can apprehend affective
properties independently of our present affective states. However, unlike
Wollheim, the solution to the puzzle provided by the proposed similarity
explanation does not explain atmospheres as properties which have their
origins in a projection from our affective states onto the world. Rather, it
solves the puzzle by indicating a similarity between phenomena which are,
in fact, distinct in kind.

To sum up, the similarity explanation provides an answer to solve
the puzzle along the following lines. Atmospheres (as a kind of affective

properties) receive the same names as moods (which are a kind of affective state) because atmospheres make things appear to us as having a similar aspect, look, and colouration as the one that we project onto the world while being in the corresponding mood.

## Conclusion

In this chapter, I have argued that moods as affective states are sharply distinct from atmospheres as affective properties. This distinctiveness thesis provided a clear answer to the question whether moods and atmospheres are, in fact, the same phenomena. I have also argued that atmospheres are apprehended by an intentional feeling and that this feeling might elicit a wide range of affective responses. Finally, I have provided an answer to the question of why despite moods and atmospheres being distinct in kind, we employ the same terms for both. I have argued that this not because of a whim of language, but rather because there is a similarity between the two phenomena.

## Notes

1 We rarely speak of atmospheres (and of affective properties) with the same terms that we use to refer to our emotions. A landscape might be calm, melancholic, cheerful, sad, etc. but not jealous, envious, ashamed, or regretful.

2 Although most philosophers have interpreted intentionality in terms of *aboutness*, arguing that while emotions are intentional states, moods lack intentionality (Searle 1983, 1), here I adopt an intentionalist view of moods. For intentionalists, rather than asking what your anxiety, elation, or irritability is about, we should be asking how each of these moods makes things *seem* to you (Crane 1998, 241; more recently Kriegel 2019).

3 Although I develop my analysis of the structure of moods via comparison with the emotions, the differences regarding their objects are sufficiently distinct to consider moods as a different state (for a different view, see Goldie 2000 and Rossi 2019).

4 The relation between emotions and their formal objects has been the object of dispute in contemporary research. Here I adopt Scheler's view according to which emotions are responses to evaluative properties (Scheler 1973, 259; Vendrell Ferran 2008, 2005).

5 Early phenomenologists are well known for defending realism and objectivism about values. What is less known is that they also embraced subjectivist positions. In this vein, though Geiger argued that evaluative properties are grasped through feeling, he also acknowledged the possibility that we project the light of our affective states onto the world. In a similar vein, Voigtländer argued that there are evaluative properties (such as impressive, imposing, etc.) which the subject impresses on the targeted objects (she names these properties: "Eindruckswerte") (Voigtländer 1910, 80). On Voigtländer's account, see Salice (forthc.) and Yaegashi (forthc).

6 For the view that emotions have three objects, see Mulligan (2015). Mulligan understands the third object in terms of a feeling character or atmosphere whose apprehension precedes the emotion. By contrast, here I endorse a different view more in line with Geiger's proposal, according to which the third

object is a colouration or feeling tone impressed onto the world and as such its apprehension does not precede the emotional experience. Put otherwise: for me, the third object is what Geiger calls "feeling tone," while Mulligan seems to suggest that the third object is comprised of what Geiger calls "feeling characters."

7  While Böhme and Schmitz seek to challenge the subject-object divide, Griffero is closer to my idea of atmospheres as properties (2014).

8  This feature is common also to evaluative properties such as beauty, unfairness, etc. For instance, for a bouquet of flowers to be beautiful, it depends on the forms, colours, the composition of the flowers that constitute it. If we change one of the flowers for another which is different in colour and form, then the bouquet might lose its beauty (for aesthetic properties, see Reicher 2010, 60).

9  In my view, these are the main models presented in the research. However, I do not pretend to be exhaustive in my discussion of models and authors. There is a prolific literature on the apprehension of such properties in music too, which I will not enter into here.

10  The origins of this view can be found in the Brentanian and early phenomenological tradition.

11  These moments exhibit specificities for each phenomenon belonging to this family. For instance, emotions are responses to evaluative properties, while moods make us more sensitive to certain evaluative properties.

12  These arguments were put forward already by Reinach (2017, 211) and Scheler (1973, 257). In contemporary research, they have been developed by Mulligan (2009) and other critics of the perceptual model (e.g., Engelsen 2018, 240).

13  As a result, the thesis that moods exhibit a distinctive intentionality has to be understood beyond the frame of representationalism (Hatzimoysis 2017).

14  The core idea here is derived from a dichotomy between states and activities of the mind which can be found in classical phenomenology (Reinach 2017, 109; Scheler 1973, 257). Here, however, I adopt contemporary terms (Mourelatos 1978, 423).

15  In fact, Lipps distinguished four main types of empathy: a) empathy of activity; b) empathy of mood; c) empathy with nature; and d) empathy into the sensuous experience of other beings (1903, 96–223).

16  For a discussion and critique of Geiger's account, see Griffero (2014, 109, 132, 134) and Vendrell Ferran (2019, 292 and 293).

17  Unlike Scheler, I consider here the feeling of affective properties and the feeling of axiological properties to refer to the same human ability but directed towards two different kinds of objects (more specifically, I consider that affective properties and axiological properties are both evaluative properties grasped through the same mechanism of feeling).

18  He refers to this feeling as "perceiving emotionally."

19  Phenomenologists used to speak of feelings as an ability (Fähigkeit) which can be cultivated and enhanced, but it can also become impaired and we might even become blind to it (see, for instance, Pfänder 1973, 54).

20  For a different analysis of the experience of atmospheres, cf. Krebs (2017).

21  I adopt here a tripartite distinction which can be found in von Hildebrand (2016, 368).

22  This is a constant puzzle in the discussion about affective properties (see, for instance, Krebs 2017, 1420 and 1423; Osborne 1968, 107).

23  In his paper on Stimmungen, Thonhauser argues that Geiger's solution is "the closest we can get to the attribution of Stimmung to objects within a psychological framework" (Thonhauser 2020). In this paper, I took this frame as a point of departure by considering that moods and atmospheres are sharply distinct.

24 To be precise, Geiger also acknowledges that the causal explanation might play a role.
25 The idea of aspect or look should not be conflated with the idea of expression. When we call a landscape sad, the way in which this landscape looks need not have anything in common with the typical bodily and facial expressions of sadness in sentient beings.

# References

Böhme, Gernot. 2001. *Aisthetik. Vorlesungen über Ästhetik als allgemeine Wahrneh-mungslehre*. München: Fink.
Böhme, Gernot. 2006. *Architektur und Atmosphäre*. München: Fink.
Crane, Tim. 1998. "Intentionality as the Mark of the Mental." In *Contemporary Issues in the Philosophy of Mind*, edited by Anthony O'Hear, 229–251. Cambridge: Cambridge University Press.
Engelsen, Soren. 2018. "Feeling Value: A Systematic Phenomenological Account of the Original Mode of Presentation of Value." *The New Yearbook for Phenomenology and Phenomenological Philosophy* 16: 231–247.
Geiger, Moritz. 1976. "Zum Problem der Stimmungseinfühlung." In M. Geiger, *Die Bedeutung der Kunst: Zugänge zu einer materialen Wertästhetik*, edited by Klaus Berger and Wolfahrt Henckmann. München: Fink.
Geiger, Moritz. 1974. *Beiträge zur Phänomenologie des Ästhetischen Genusses*. Tübingen: Niemeyer.
Goldie, Peter. 2000. *The Emotions: A Philosophical Exploration*. Oxford: Oxford University Press.
Griffero, Tonino. 2014. *Atmospheres: Aesthetics of Emotional Spaces*. Farnham: Ashgate.
Griffero, Tonino. 2017. *Quasi-things: The Paradigm of Atmospheres*. Albany: SUNY Press.
Haas, Willy. 1910. *Über Echtheit und Unechtheit von Gefühlen*. Nurnberg: Benedikt Hilz.
Hatzimoysis, Anthony. 2017. "Representationalism and the Intentionality of Moods." *Philosophia* 45, no. 4: 1515–1526.
Hepburn, Ronald W. 1968. "Emotions and Emotional Qualities: Some Attempts at Analysis." In *Aesthetics in the Modern World*, edited by Harold Osborne, 81–93. New York: Weybright and Talley.
Kenny, Anthony. 1963. *Action, Emotion and Will*. London: Routledge & Kegan Paul.
Krebs, Angelika. 2017. "*Stimmung*: From Mood to Atmosphere." *Philosophia* 45, no. 4: 1419–1436.
Kriegel, Uriah. 2019. "The Intentional Structure of Moods." *Philosophers' Imprint* 19, no. 49: 1–20.
Lipps, Theodor. 1903. *Ästhetik: Psychologie des Schönen und der Kunst*. Hamburg-Leipzig: Voss.
Morris-Jones, Huw. 1968. "The Language of Feeling." In *Aesthetics in the Modern World*, edited by Harold Osborne, 194–204. New York: Weybright and Talley.
Mourelatos, Alexander. 1978. "Event, Processes, and States." *Linguistics and Philosophy* 2, no. 3: 415–434.
Mulligan, Kevin. 2009. "On Being Struck by Value." In: *Leben mit Gefühlen*, edited by Barbara Merker, 141–163. Paderborn: Mentis.

Mulligan, Kevin. 2015. "Secondary Meaning, Paraphraseability and Pictures: From Hofmannsthal to Wittgenstein." In *L'expression des Émotions: Mélanges Dédiés à Patrizia Lombardo*, edited by M. Rueff and J. Zanetta. Geneva: University of Geneva Press.

Osborne, Harold. 1968. "The Quality of Feeling in Art." In *Aesthetics in the Modern World*, edited by Harold Osborne, 105–124. New York: Weybright and Talley.

Pfänder, Alexander.1973. *Ethik in kurzer Darstellung*. München: Fink.

Price, Carolyn. 2006. "Affect Without Object: Moods and Objectless Emotions." *European Journal of Analytic Philosophy* 2, no. 1: 49–68.

Reicher, Maria E. 2010. *Einführung in die philosophische Ästhetik*. Darmstadt: WBG.

Reinach, Adolf. 2017. *Three Texts on Ethics*. Munich: Philosophia.

Rossi, Mauro. 2019. "A Perceptual Theory of Moods." *Synthese*. https://doi.org/10.1007/s11229-019-02513-1

Salice, Alessandro. (Forthcoming). "Else Voigtländer on Social Self-Feelings." In *Else Voigtländer: Self, Emotion and Sociality*, edited by İngrid Vendrell Ferran. Cham: Springer.

Scheler, Max. 1973. *Formalism in Ethics and Non-formal Ethics of Values*. Evanston: Northwestern University Press.

Schloßberger, Matthias. 2019. "Phänomenologie der Naturerfahrung. Einfühlung und Einsfühlung mit der Natur bei Geiger und Scheler." In *Natur und Kosmos*, edited by Hans Reiner Sepp. 1–21. Nordhausen: Bautz.

Schmitz, Hermann. 2005. *System der Philosophie III*. Bonn: Bouvier.

Schmitz, Hermann. 2008. "Gefühle als Atmosphären." In *Atmosphären im Alltag*, edited by S. Debus, 260–280. Hannover: Psychiatrie-Verlag.

Searle, John. 1983. *Intentionality*. Cambridge: Cambridge University Press.

Sizer, Laura. 2000. "Towards a Computational Theory of Mood." *British Journal of Philosophy of Science* 51, no.4: 743–769.

Solomon, Robert. 1993. *The Passions: Emotions and the Meaning of Life*. Indianapolis: Hackett.

Tappolet, Christine. 2000. *Emotions et Valeurs*. Paris: Presses universitaires de France.

Thonhauser, Gerhard. 2020. "Beyond Mood and Atmosphere: A Conceptual History of the Term Stimmung." *Philosophia* 49: 1247–1265.

Vendrell Ferran, İngrid. 2008. *Die Emotionen. Gefühle in der Realistischen Phänomenologie*. Berlin: Akademie.

Vendrell Ferran, İngrid. 2019. "Geiger and Wollheim on Expressive Properties and Expressive Perception." *Studi di astetica* 2: 281–304.

Voigtländer, Else. 1910. *Vom Selbstgefühl*. Leipzig: Voigtländer.

Von Hildebrand, Dietrich. 2016. *Aesthetics I*. Steubenville, OH: Hildebrand Project.

Wollheim, Richard. 1987. *Painting as an Art*. Cambridge, MA: Harvard University Press.

Yaegashi, Toru. (Forthcoming). "Erotic Love and the Value of the Beloved." In *Else Voigtländer: Self, Emotion and Sociality*, edited by İngrid Vendrell Ferran. Cham: Springer.

# Part II

# Psychopathological atmospheres

# 4 Atmospheres of anxiety

## The case of Covid-19

*Dylan Trigg*

## Introduction

The emergence of the infectious disease known as Covid-19 has caused widespread death and illness, economic unrest, and global uncertainty, the impact and extent of which remains presently unknown. Families have been destroyed, jobs lost, and healthcare systems overwhelmed. Throughout this, there has also been an expediential growth in anxiety across different populations. Whether it be anxiety relating to health, the economy, relationships, or more generalised anxiety directed towards future uncertainties, there can be little doubt that Covid-19 is having and will continue to have a sharp impact on mental health (cf. Elliott 2020; McKie 2020; Riberio 2020). Yet the anxiety produced by Covid-19 is not only an affective state experienced by individuals, it is also something that is extended in the everyday world as part of a general *atmosphere*.

The idea of anxiety as a type of atmosphere is in contrast to how the emotion has commonly been thought about (cf. Trigg 2016). Whereas atmosphere has the connotation of an affective force that is diffused in the world, we tend to think of anxiety as an emotional state experienced by individuals alone. Indeed, this tendency to think of anxiety in individualistic terms is already evident in the history of phenomenology. The case of Heidegger is emblematic of this approach (cf. Bergo 2020). For Heidegger, anxiety is the philosophical mood *par excellence*, which strips everyday life of its aura of familiarity and reveals the contingent foundations upon which meaning is constructed (Heidegger 2008). The result of this rupture is that we feel "ill-at-home" in the world, as the world reveals itself in its strangeness (Heidegger 1977, p. 111).

Yet this is not a movement of negativity for Heidegger, but instead an opportunity for us to redefine our relationship to the world and to others. As he sees it, our inauthentic and largely non-anxious rapport with others consists of an absorption in the generalised and anonymous voice of the many. It is not that *I* have such and such a belief about the world, but that *they* do (cf. Heidegger 2008). *One* feels, *one* thinks, and *one* acts in a certain way because it is what *one* just does. This complicity with an anonymous

DOI: 10.4324/9781003131298-5

one is what anxiety shatters. Given this, the realisation of Heidegger's project consists of employing anxiety as the means for human existence to undergo a transition from absorption in the anonymous masses to a state of individuation, marked by an affirmation of both death and anxiety as *my own*. To this end, the value of anxiety in Heidegger is framed at all times by its power to individualise the subject.

While Heidegger's account of anxiety has proved highly influential within phenomenology, philosophy, and the humanities more generally, such an account nevertheless tends to prioritise a moment of self-realisation within an individual. As a result, what the account neglects is the way anxiety is distributed through the world and embodied in other people and in things themselves. In this chapter, I would like to propose a challenge to the Heideggerian tradition of treating anxiety as an opportunity for self-transformation by conceiving it as an *atmosphere*.

As is evident in this collection, the concept of atmosphere has gained a significant amount of attention in several disciplines, especially philosophy, human geography, and literary studies (Böhme 2017; Griffero 2014; 2017). Notwithstanding the diversity characterising current research on atmospheres, there are a number of salient themes common to their conceptualisation; namely, atmospheres an affective yet indeterminate phenomena that are grasped pre-reflectively, felt corporeally, and given expression through material bodies (cf. Trigg 2020). Yet despite these common themes, the concept of atmosphere is also inherently ambiguous. This level of ambiguity is reflected in questions that populate the literature on atmospheres concerning the relationship between atmospheres and moods, the question of where atmospheres derive from, and the extent to which atmospheres can be shared (cf. Griffero 2014).

Atmosphere's ambiguity is also captured in an experiential sense, as when we talk about certain rooms as having an "eerie," "tense," or "buoyant" atmosphere. In a similar measure, we frequently talk about political events as having an atmosphere of a relief in the context of an election or otherwise suffering from a fraught atmosphere in the context of on-going discussions. Terminologies such as these, while implicitly understandable, nevertheless belie a complex substructure of meanings, which are difficult to pin down in unequivocal terms, and, which issue a challenge to the idea of affectivity and emotion being a cognitive or even subjective phenomena alone (cf. Sumartojo and Pink 2019). Yet far from an incidental or contingent aspect, the ambiguity peculiar to the concept of atmosphere is what generates its specificity. Indeed, it is because the concept of atmosphere extends beyond the remit of individual emotion, embedding itself in material structures and being enacted through cultural practices, that it has assumed an influence beyond that of any single discipline.

In the current investigation, my core point of departure is that the concept of an atmosphere can play a powerful role in accounting for (i) how anxiety is distributed through the world and (ii) how anxiety can institute and express itself in specific things without being reducible to those things—claims that I will unpack in the course of this chapter. The test study for this hypothesis is the anxiety tied up with the current and on-going coronavirus crisis. As indicated above, the anxiety interwoven with Covid-19 is multidimensional and complex, and it is important at the outset to issue a disclaimer with respect to my area of focus. What interests me here is the affective experience of being under lockdown in an urban or densely populated environment. As such, what I do *not* focus on is the lived experience of illness that comes with coronavirus as a specific kind of disease. Nor do I explicitly consider the societal, ideological, and political aspects of Covid-19. Rather, I am concerned with the ways Covid-19 augments and shapes everyday experience, and how this transformation of the everyday can impact our broader experience of the world.

Moreover, the formulation of anxiety in this study is delimited to a *generalised* form. I emphasise that the mode of anxiety involved in Covid-19 is generalised because it is important to qualify that the anxiety at work in this atmosphere does not exhaust or incorporate every articulation of anxiety. To be sure, while certain types of anxiety might be especially susceptible to the affective register of Covid-19 (to think here of anxiety accompanying agoraphobia, panic, and obsessive compulsive disorder), these are specific instances of anxiety, which are conceptually and thematically different from the generalised and non-pathological anxiety peculiar to existence during and after lockdown (though, equally, there is nothing to prevent there being an overlap between panic and generalised anxiety, but that is not the focus of the current chapter).

The concern of the current chapter is how generalised anxiety shapes the practice of everyday life. As I use the term, generalised anxiety is a mode of anxiety that is inherently atmospheric insofar as it alters our fundamental relationship to the world while also manifesting itself in specific and localised issues (cf. Kluger 2020). To this extent, the current formulation of anxiety is consistent with my previous account of the emotion in *Topophobia: a Phenomenology of Anxiety* (Trigg 2016), in that I regard anxiety as having both localised articulations (i.e., a specific phobia of a bridge or a social situation) while also permeating the world in a diffused and non-specific way. Furthermore, while I find aspects of the canonical phenomenological treatment of anxiety beneficial—especially the Heideggerian analysis of uncanniness—I nevertheless dissent from the conventional distinction between fear and anxiety, together with the adjoining commitment to anxiety being an intentional state without an object.

The plan for this study is threefold. First, I consider the intentional structure of an atmosphere, giving special attention to the way an atmosphere generates a specific affective style, which is given expression in both a diffused and singular sense. Second, I consider one of the salient themes of Covid-19 anxiety; namely staying at home and leaving home. As I argue, Covid-19 is not a discrete and localisable phenomenon, but instead a force that redefines boundaries and reconfigures our experience, interpretation, and understanding of the outside world. Finally, I consider how the lived body is augmented in and through the lens of coronavirus. Here, I posit that one of the key aspects of Covid-19 is that it thematises the body in its thing-like status, in turn, issuing a challenge to the idea of the body as irreducibly "one's own." I end by consolidating the role atmospheres play in synthesising these elements together.

## Urban atmospheres

It is worth beginning here by framing the ways an atmosphere is diffused in an urban environment. The question of whether such an atmosphere would be felt in a non-urban environment is a decisive one, yet which lies outside the scope of this chapter. Nevertheless, given that Covid-19 as a particular kind of disease thrives upon densely populated spaces, it is *prima facie* evident that such spaces are more commonly found in urban environments. At the same time, an urban environment is not a pre-determined or ahistorical substance unaffected by the people who transit through it. Rather, peoples and their specific cultures shape the presentation of an atmosphere just as atmospheres, in turn, colour and influence our moods. This is as much evident in terms of how individual moods generate a different interpretation of an atmosphere as it is the way entire cultures respond to regulations enforced by lockdowns (cf. Overy 2020).

My focus on an urban environment is predicated on the sense that such an environment captures the atmosphere of anxiety in several key respects. The first point to note is that a city is never devoid of an atmosphere. Even an innocuous city characterised by homogenous and pre-fabricated structures—a city of interchangeable megastores, lifeless streets, and an endless parade of concrete—even such a city carries with it a thick atmosphere. True, the atmosphere in question may be presented in pernicious terms as a threat to the notion of a city as having a plentiful and heterogeneous character, but such resistance simply attests to the weight of a homogenous atmosphere as having its own singular quality. More often, however, when we talk about the atmosphere of a city, then we do so in terms of a tonality that permeates and gives character to a place, or by what Merleau-Ponty calls "a latent sense, diffused throughout the landscape or the town, that we uncover in a specific evidentness without having to define it" (Merleau-Ponty 2012, 294). Consider here how we speak of the romantic atmosphere of Paris, the frantic atmosphere of New York, or the imperial atmosphere of Vienna. What these terms tend to denote is a constellation of historical, cultural, political, and aesthetic structures cohering into the same orbit. These constellations are not affective elements staged together as an image—though of course they are often manipulated and capitalised upon for political or commercial gain—but instead emerge as an affective presence, which gives a city its sense of place (cf. Böhme 2017).

Notably, the idea of an urban atmosphere is not something we can understand in abstraction. Rather, atmospheres are understood through being grasped in a bodily sense. To talk about the vibrant atmosphere of New York is to speak about a certain articulation of the city characterised by a unique set of sensual, aesthetic, cultural, and material properties. Atmospheres, thus, appear for us in a multidimensional way. They are spaces characterised as murky, piercing, soft, threatening, inviting, eerie, charmed, and so forth. Likewise, to sense an atmosphere as anxious is not to make a set of abstract deductions about a given situation. More primordial than this, to be anxious is to be grasped in a pre-reflective sense by the bodily experience of anxiety (cf. Trigg 2016). Moreover, we grasp such states not as a set of discrete parts to be understood in representational terms, but as a unitary phenomenon felt immediately in and through the lived body in its relation to the surrounding world. Indeed, it is often the case that we simply grasp the specific atmosphere of a place in an instant without having to think about it in abstraction.

What does it mean to speak in specific terms about Covid-19 as an atmosphere of anxiety that permeates an urban environment? As I will try to spell out in the following sections, it means giving some specificity to how this atmosphere affects and shapes cultural and bodily practices, spatial configurations, and a generative way-of-being that marks our relationship to urban spatiality as a whole. On this point, the first task of this chapter is to pose an initial question; namely, from where does anxiety emerge, and, where is it located? In asking this question, it becomes necessary here to carefully consider the intentional structure of an atmosphere of anxiety, which we will now do.

## The intentional structure of Covid-19 anxiety

Ordinarily, we understand intentionality as a mental act directed towards discrete objects, whether real or not (Husserl 2001; Merleau-Ponty 2012). Thus, in classical phenomenology, intentionality refers to how our mental states are always *about* something. But this quality of being about something is not simply a mental or intellectual act; it is also embodied, affective, and often unconscious (Merleau-Ponty 2012). Seen in this respect, intentionality extends beyond that of a mental act and instead frames that way in which we engage with the world in a prereflective and prepersonal sense. Thus, to experience the world as an anxious place is not to intellectually posit a certain risk with respect to our relationship to the world; rather, it is to be attuned to the world as a site of meaning that is diffused as a general style.

Merleau-Ponty speaks in this light of an "operative intentionality," which "establishes the natural and pre-predicative unity of the world and of our life" (Merleau-Ponty 2012, lxxxii). What this means is that beneath the level of abstract thought, a more primary mode of intentionality is at work, "appears in our desires, our evaluations, and our landscape more clearly than it does in objective knowledge" (lxxxii). This level of intentionality incorporates non-thetic modes of consciousness, which can ultimately explain how meaning is posited in a non-intellectual way and, thus, can provide a key role in accounting for the intentional structure of an atmosphere, as Merleau-Ponty has it, operative intentionality is "already at work prior to every thesis and every judgment" (453).

To get a sense of this with respect to the structure of an atmosphere, consider here the variation of anxieties associated with Covid-19 and its lockdown. These anxieties range from issues concerning the absence of a social life, to concerns over finances, and to a sense of being ill-at-ease while being outside during lockdown (cf. Kurth 2020; McKie 2020). In each of these variations, the lived experience of anxiety is irreducible to a singular thematic object, but is instead diffused through the environment in a porous and non-containable way. When we think here of the atmosphere of anxiety under Covid-19 lockdown, then it would be difficult to pinpoint with accuracy where the anxiety is located. Specific phenomenal features may well present themselves in a more focal way than others—a discomfort with being outside, a concern over the future, the onset of a new cough, etc.—but aspects such as these are expressions of an already existing atmosphere rather than containable and delimited tokens of anxiety. In this respect, instead of being directed towards a discrete thing, the intentional direction of an atmosphere is diffused through an environment in a multiplicity of ways and on an operative and reflective level. For this reason, an atmosphere does not present itself perceptually as a containable object in the way that a table or a chair does. Rather, there is a porous and dynamic quality to an atmosphere, insofar as it is already established as a field of meaning before it is understood, as such.

To give some flesh to these concepts, let us take an example that is relevant to the era of Covid-19; namely, being in a supermarket. The example is emblematic of a set of tensions and anxieties peculiar to our current situation. In effect, what is notable about the condition of living under lockdown is the transformation of the everyday and the prosaic into a site of anxiety and tension, which is expressive of more pervasive concerns. Indeed, it is precise because a supermarket is ostensibly devoid of significant affective value that its presentation as a flashpoint of anxiety merits remark.

There are several movements of anxiety to consider here. First, there is the phenomenon of panic buying. During the initial stages of the lockdown, media reports were saturated with accounts of shortages in the supply chain, especially concerning the availability of hand sanitiser and toilet paper, the consumption and, in turn, shortage of which became illustrative of a new age of anxiety (cf. Wilson 2020). How to understand this behaviour? On the surface, the logic of panic buying seems to be motivated by a desire to generate a sense of control in a situation where uncertainty is rampant (Yuen et al., 2020). Yet the onset of panic buying is also propelled by the notion of anxiety as a contagious phenomenon, which emerges from a more global atmosphere of unrest. After all, the stockpiling of supplies does not derive from a mode of rational reflection on the needs of an individual or a group; rather, it is given force through an ambient tonality that is felt in the streets and in the comportment of people.

This atmosphere of anxiety is also articulated in the supermarket itself. Under lockdown, the supermarket has become a place rich with multiple forms of anxiety. It is a place that one must venture into in a focused and precise way, careful to both avoid getting too close to other people but also ensuring to avoid unnecessarily touching surfaces that may be contaminated; it is also a place where there is a sense of being exposed to danger while carrying out that most primitive of tasks—gathering food. As such, the supermarket is an "essential" place both in terms of sociological and economic value, but also in terms of providing the basic fundaments of existence, and thus for many people an unavoidable excursion.

Each of these aspects does not necessarily hold greater sway over another aspect; rather, each expression of anxiety emerges in a dynamic and fluid way. The atmosphere of anxiety peculiar to the supermarket effectively seeps through the whole place, infecting both people and things within its sphere of influence. As Böhme writes, "We are unsure where they are. They seem to fill the space with a certain tone of feeling like a haze" (Böhme 1993, 114). As omnipresent within a given space, we find ourselves "gripped" by an atmosphere, affected by it insofar as it gets under our skin. In this respect, the atmosphere of anxiety within a supermarket exacts a power over the individuals within it, and unless we have cultivated a method of "tuning it out," then we remain, for better or worse, swayed by the affective tonality at work.

Understood in this way, an atmosphere permeates things within a specific *style*. Here, style denotes the broadly Merleau-Pontean sense of an affective constancy within a given atmosphere (cf. Merleau-Ponty 2012). What this means is that an atmosphere is held together by dint of an affective tone that permeates bodies—human and non-human—within its field of influence. Of this point, Merleau-Ponty speaks aptly of Paris as having a "certain style" from which specific phenomena features gain their affective tonality, writing that "[j]ust as a human being manifests the same affective essence in his hand gestures, his gait, and the sound of his voice, each explicit perception in my journey through Paris—the cafés, the faces, the poplars along the quays, the bends of the Seine—is cut out of the total being of Paris, and only serves to confirm a certain style or a certain sense of Paris" (294). To this extent, style is what gives an atmosphere its affective "momentum." An atmosphere does not surround and encompass us in a homogenous or free-floating way. Rather, there are multiple points of convergence, where we feel "closer" to or "further" from the affective centre of an atmosphere. This is especially evident during the Covid-19 lockdown. The levels of anxiety experienced by individuals (and groups) during the lockdown are not experienced as inert units of emotion unwavering in their presence; rather, anxiety flares up in waves of disquiet before receding again.

Furthermore, these expressions of anxiety hinge upon a contextual world, in which Covid-19 is in some sense omnipresent and, thus, inescapable, both spatially and temporally. The inescapable quality of Covid-19 is captured in the sense of an atmosphere that structures movement, thought, and action on both an implicit and explicit level. So long as it exists, and so long as the Earth remains a human one, then it cannot be contained in strict terms. As such, the world now presents itself, to use a term recently employed by Matthew Ratcliffe and Ian James Kidd, as a "Covidworld" (Ratcliffe and James Kidd 2020). Notwithstanding their critical assessment of the role lockdown itself plays in mitigating coronavirus—which is another topic altogether—what is notable about Ratcliffe and Kidd's essay is the usage of the term *world* in describing the presence of Covid-19 as a feature of contemporary existence. The term "world" here designates an entire system of meanings, which generates a context from which things matter to us, as they put it: "Consider how, during the course of daily life, some things appear more *salient* than others—they light up for us, stand out, grab our attention. These things also *matter* to us in different ways: maybe they excite us, threaten us, comfort us, draw us in, or repel us" (Ibid.).

This is also borne out by the sense that anxiety during lockdown has specific articulations of intensity, for example, in being outside, on public transport, or in amongst a crowd of people (to say nothing here of digital and online space, which constitutes an extension and often an amplification of offline atmospheres). These configurations of life are not innocuous relationships we assume to the world, but are instead illuminated within the realm of Covidworld. In each case, intentional orientation towards coronavirus is not directed towards a singular thing—that is, towards a discrete disease known as "Covid-19"—but instead orientated in a diverse and divergent series of ways, each of which is unified through their diffusion in a global atmosphere.

This account so far raises a critical question; namely, can one inoculate oneself from a given and prevalent atmosphere? To put it another way, if an atmosphere is diffused throughout the world on both an operative or latent level, but also at a reflective level, then how can we escape being under the sway of an atmosphere? After all, it is manifestly the case that not everyone living under lockdown and in the midst of Covid-19 is exposed to the same levels of anxiety. Here, I am not only thinking of people maintaining their plans and habits *in spite of* Covid-19 (as, for example, people who willingly ignore the lockdown) nor am I necessarily thinking of Covid-19 deniers (whose very denial of the disease, in turn, affirms it); what is also at stake here is the case of individuals who are purportedly ignorant of the reality of coronavirus, and thus, effectively live outside of its atmospheric influence.

One response to explain this phenomenon is through the deployment of what might be called a *counter-atmosphere*. If a collection of individuals or an entire group is implicitly committed to the creation and maintenance of an atmosphere, which is in some sense *their own*, then it is feasible that such an atmosphere should run in parallel to a dominant atmosphere without there being considerable overlap. The same structure is no less true of individual moods. When we are deeply immersed in a mood—depression or anxiety, for example—then the world is illuminated as a depressed or anxious world. This interpretive immersion seldom leaves space for the co-existence of alternative moods, especially those that are affectively incompatible with depression or anxiety. The difference with the atmosphere of Covid-19, however, is that instead of being an idiosyncratic mood we find ourselves in, the disease is instead a global phenomenon against which specific injunctions have been placed on how we ought to act and move in the world. As such, a level of ignorance cultivated in a counter-atmosphere is only tenable on a short-term basis; at some point, the restrictions orchestrated by governments will invariably infringe upon these groups, and the counter-atmospheres that form a respite from Covid-19 will be diminished if not extinguished by the prevalent and dominant atmosphere of our era.

## Stay at home

From a general tonality in the air, we move now to a series of specific articulations, which characterise the structure and nature of Covid-19 as an atmosphere. One way the atmosphere of Covid-19 is given a dominant expression is through the injunction to *stay at home*. This slogan, which was repeated throughout press briefings and throughout the media more broadly, establishes a sharp delineation of the urban environment that was hitherto largely inconceivable for the majority of people. The prospect of having to stay at home was unthinkable not only as an empirical issue, but also in terms of how we ordinarily understand our relationship to the surrounding world. For the most part, we take for granted the idea of home not only as a discrete place in the world, but also as a mode of being-in-the-world more generally. To be at home in the world means being situated in a milieu that is framed by a movement of possibility and potentiality. We are "at home" in the world, that is to say, insofar as the world continues to renew itself in a dynamic and spontaneous way, revealing itself as a nexus of meaningful relations (cf. Trigg 2012; 2016).

On an experiential level, this sense of the world as *homely* presents itself with an atmosphere of constancy. Generally speaking, the world appears to us with an air of familiarity, such that we always have a sense of where we are even if lost. Over time, people and places intertwine, and we become attached to neighbourhoods, cities, and entire countries until they sediment themselves in us as part of the fabric of identity. Understood in this way, to leave the home and venture forth into the world is to frame the physical home as a point of departure rather than a zone of termination. Home, conceptualised as a set of relations and values, is, thus, not delimited to a site but is instead formulated as a sense of being-at-home within the world more generally (cf. Merleau-Ponty 2012). As such, "home" indexes as much an environment in the world as it does a specific kind of implicit (and atmospheric) relationship we have to that environment, from which our actions, emotions, and intentions emerge.

Against this atmosphere of tacit confidence and taken-for-granted certainty, the injunction to *stay at home* establishes a radically different idea of home. Instead of being a concept diffused in the world, the notion of "home" during Covid-19 is delimited and reduced to a physical dwelling, where one is obliged—indeed, *obligated*—to be. The implication is that home—while possibly mitigating some of the dangers associated with Covid-19—nevertheless becomes less a place of sanctuary and more a site of constriction and tedium, which seals the dweller from the world and fragments a meaningful sense of being-at-home in the world. On this point, Kevin Aho writes aptly, "We see now that there are different ways that we can die during a pandemic. We can, of course, perish in a hospital bed due to coronavirus. But we can also die an ontological death when we lose our hold on things, when we lose our self-understanding and are unable-to-be because we are cut off from purposive, meaning-giving involvement in the world" (Aho 2020, 15).

Aho's point underscores the relational and meaning-laden value associated with the concept of home. For many people, the home that is now the centre of life is not quite the place it once was. As sealed from the world, home is not a porous concept that spills over into the world; rather, it is an endpoint that fragments referential meaning and engenders a sense of the world as constricted and compressed. Moreover, the very phenomenology of the home itself has changed; now, home is understood as a site of potential contamination that must be disinfected and sterilised before it can open itself up to being dwelt in. The result is an articulation of the home as that which is simultaneously homely and unhomely, personal and impersonal, and familiar and unfamiliar in the same measure (cf. Aho 2020; Trigg 2016).

Notably, this onus on staying at home has a series of critical consequences for our relationship with the outside. In typical circumstances, the boundary between inside and outside is porous and dynamic. We move from the home with an implicit trust that the outside world is neither an affront to our existence nor in sharp contrast to it. In this respect, distance and movement are not understood as abstract grids of references that are mapped out in advance; rather, they are textures and contours of a living spatiality, which are grasped in an affective sense, as Merleau-Ponty writes: "We know a movement and a moving something without any consciousness of the objective positions, just as we know a distant object and its true size without any interpretation ... Movement is a modulation of an already familiar milieu" (Merleau-Ponty 2012, 288).

Merleau-Ponty's point that movement is predicated on an already familiar world is given vivid expression during the initial phases of the lockdown. Here, the already familiar world is presented through an unfamiliar lens where movement becomes tentative and framed at all times by a gesture of constriction. Moving from one point to another—from the home to the supermarket—becomes less a mindless activity carried out as part of the fabric of everyday life and more a series of movements, which demand that we restructure our relationship to the world. In effect, spatiality as a constant flow of emergent properties has become a dissected cluster of habitable and uninhabitable zones. The result is a double-bind, in which both public and private space are equally infected by the atmosphere of Covid-19, as one individual says in an interview with Time magazine, "I'm concerned about going into public, but now I'm also concerned about how long I can [last] without going out" (Ducharme 2020).

## A touch of anxiety

Alongside this unnatural and disturbed relationship to the outside, the materiality of the world itself assumes a strange tonality. The world, as it is commonly understood and experienced, is suddenly suspended. It is true, on the surface, it appears as though nothing significant has changed; buildings remain intact, people can be seen casually milling about, and some shops are even open for business. The world has not collapsed and the end, if approaching, is not here yet. Yet in the midst of this apparent normality, a deadly pandemic has taken hold, which creates a series of subtle shifts in our everyday experience. While still available to us, the outside world is nevertheless experienced through a hyper-sensitised lens, in which tactility as much as vision plays a dominant role. Consider here how surfaces that were once invisible and innocuous have now become charged with a sense of being sites of disease. Everyday objects—phones, door knobs, elevator buttons, etc.—are also altered from familiar tokens of everydayness residing in the background to things charged with an aura of danger.

In no uncertain terms, Covid-19 has issued a stark challenge to the idea that human experience as being strictly ocularcentric. In place of this notion, it is touch that has become our primary mode of being-in-the-world. The injunction to avoid touching one's face, and surfaces more generally, reinstates—albeit in a negative mode—the porous interplay between ourselves and the world. We are not discrete subjects gazing upon an otherwise blank world; rather, to put it again in Merleau-Pontean terms, just as we touch the world with our sensing organs, so the world touches us back (Merleau-Ponty 1968). Only now, the world that is reversing our touch is a world marked by disease and harm. As such, our relationship to the world is one that has to be kept at arm's length; instead of greeting people with our entire bodies, we have had to contrive novel ways to interact with people without spreading the disease. And instead of freely touching the world around us, we have to exercise caution about which surfaces it is necessary to engage with in order to perform basic functions. As fundamentally altered, the world and its arrangement of material things protrude into our perceptual experience, thus becoming thematised in their strangeness.

True to the nature of an atmosphere, this permeation of anxiety is not localised to the lockdown itself, but instead pours into life after lockdown as a temporal expression of anxiety. "Some people," so Lily Brown, director of the Center for the Treatment and Study of Anxiety at the University of Pennsylvania's Perelman School of Medicine remarks, "are anxious because they have a 'lurking fear' of catching or spreading COVID-19 ... while others have fallen out of practice socializing and are finding it difficult to resume" (Ducharme 2020). Thus, we see that the hazy quality of an atmosphere, which Böhme speaks of, is at once spatial and temporal (Böhme 1993). Atmospheres do not reach a neat endpoint in tandem with a sequence of calendric dates; rather, they spill into the present like vapour trails from the past.

Consider how atmospheres can often linger under our skin long after we have left the place or situation from where that atmosphere derived. This is evident as much on a subjective and personal scale as it is on a cultural and political level, as when we talk about the atmosphere of one decade seeping into the beginning of the following decade. Likewise, the anxiety associated with Covid-19 marks a "lurking fear" that continues to affect our behaviour, thoughts, and actions in number of ways. As Lily Brown remarks above, this fear is captured as an epistemic gap in our knowledge of the world and as a set of bodily practices that have been attuned to a climate of tension and which now require re-training in order to adjust to life after lockdown.

The accumulative result of these new dynamics is that dwelling in the world has been put out-of-joint, such that for many of us, a sense of being "at-home" is now experienced in terms of being *ill-at-home*, to maintain the Heideggerian language (Heidegger 2008). Being ill-at-home in the world means being confronted with a world, in which the meaning underpinning actions, intentions, and thoughts have fragmented. Into this fragmented scene, things no longer assume the value they once did; the everyday itself as a nexus of relational meanings loses referential value and, as a result, a sense of anxiety permeates much of life. Central to this permeation of anxiety is the role the body plays in given expressive form to an atmosphere. We have already touched upon this with the mention of touch; but more needs to be said on the issue of the body as mediator and object of anxiety.

## The thinglike body

In "normal" life, we generally take our bodies in a pregiven and taken-for-granted way. Like the homes we dwell in, we have a trust in our bodies that provides a sense of continuity over space and time. Moving in and through the world, we do so with a tacit sense of our bodies as generating a feeling of directional, affective, and intersubjective awareness. Meeting other people, we have an implicit sense of how to conduct ourselves in proximity to other bodies. Distance and proximity are not spaces measured in abstraction, but rather degrees of intentional awareness we grasp from an experiential perspective. Phenomenology has provided an abundance of attention to this modality of embodied life; it is a body that is engaged in the world in a fundamentally affirmative way; it is a body that relates to the world in the form of an *I can* rather than an *I cannot*; it is a body that is intertwined with other bodies in a fluid and dynamic way; and it is, above all, a body that is "one's own" (cf. Merleau-Ponty 2012).

Within the research on atmospheres, such a body has assumed a critical role. Consider here how atmospheres are not only extended into the world, but are also grasped in and through the body. In the language of Schmitz, the body as felt is extended into the world insofar as it is constantly affected by the world (Schmitz 2011). The body is not an atomic entity, but an opening that is co-constituted by the spaces we inhabit and dwell in. Accordingly, just as spaces expand and contract with different affective structures so too do our bodies. This is especially clear in the case of the anxiety pertaining to coronavirus. The atmosphere of anxiety surrounding Covid-19 is not an amorphous affective force lacking direction; rather, it assumes a specific and amplified expression in certain situations and environments. As we have seen above, it is in and through urban space that the atmospheric structure of Covid-19 is given its clearest expression. Thus, nearer to the scene of anxiety, we tend to feel our bodies tightening up and our air becoming constricted. Whereas in a place of repose, we feel our bodies adjusting to the environment and, in turn, gradually exhaling.

Yet the embodied experience of Covid-19 does not simply concern a body that has been infected by a disease; more complex than this, the disease transforms the lived experience of one's own body more generally into an agent of anxiety whether a person is infected or not. This articulation of anxiety through the body has at least two aspects to it; first, in terms of one's own relation to their body, and, second in terms of one's relation to other bodies.

To begin with the first figuration, one of the striking aspects of the Covid-19 pandemic is the modification of the body from the centre of intimacy and familiarity to a site of suspicion and otherness; it is a body that is not only at risk of becoming diseased but also of being a source of alienation. One way this manifests itself is in terms of the body becoming objectified as a potential carrier of the disease. Much of the media narratives concerning Covid-19 focus on the elevation of anxiety in the population, with a specific focus on a heightened attention to changes in the felt experience of the body. Whereas the body is ordinarily a tacit presence in everyday life, in the age of Covid-19, signs and symptoms emanating from the body acquire a halo of meaning usually reserved for periods of illness and injury. A first-person report from the *Washington Post* captures this amplification of meaning vividly:

A thermometer sits on the window sill of my bathroom, right next to the toilet, so every time I go to the bathroom, I can take my temperature. I've been feeling like I have a low grade fever for weeks, and these days, a fever isn't just a fever. It's a signal you may have the coronavirus. And so I take my temperature about eight times a day to see if my fever has risen.

(Chesler 2020)

What is notable about this passage, and many other passages in the media that echo this sentiment, is that instead of being a nexus of lived meanings, the lived body is now reduced to a *Körper*, a thinglike body that has become foregrounded in its biological and anonymous materiality (cf. Trigg 2019). As it is understood in classical terms, the thinglike nature of the body is the dimension of bodily life that materialises when the body as a physical thing is foregrounded through pain, illness, fatigue, psychopathologies, ageing, and so forth. Thus, in moments of sickness, the body ceases to be an implicit structure and is instead presented to us a focal point of perception, which can disturb our sense of selfhood. Likewise, catching sight of our bodies as having aged, we tend to experience a gap between who we think we are and our bodies which, as it were, have trailed off on their own. In each case, the body is rendered a thing that we observe and monitor for further changes, and which we have little or no power over.

The same structure is no less true in the case of Covid-19. The Covid-19 body is not only an "ill" body in the sense of being infected by a disease, it is also "ill" insofar as it becomes a site of suspicion, generating a hypochondriac if not paranoid relationship to the body's materiality, as Kevin Aho writes in his incisive essay on the uncanniness of coronavirus, "My hands have become eerily conspicuous, like strange objects that I am only contingently connected to. Worried about contracting the virus, I no longer reach effortlessly for the doorknob or the cell phone; nor do I extend my hand in a warm greeting when a friend comes by" (Aho 2020, 8). Aho draws our attention here to the manner in which the body has become largely mediated through an atmosphere of anxiety. It is not that the hand as a discrete organ has become an object of anxiety in and of itself; rather, the hand gives expression to an anxiety that has already been instituted by the onset of Covid-19. In correspondence, sensations which previously had a contextual meaning within relationship to the everyday—headaches, tiredness, etc.—all now point towards a singular horizon; Covid-19. At the heart of this paranoid logic is the uncertainty of what is at stake in each of the body's processes, responses, and symptoms. The body that is presented to us in the face of Covid-19 is, thus, in large and unknown and unknowable body; it is a body that is ambiguous not only in terms of being both a thing and a centre of perception, but also in terms of being both mine and not-mine concurrently.

## Other bodies

It is not only our own bodies that undergo a shift in their perceptual and affective structure, but also our relation to *other bodies*. In normal circumstances, our communication with others is orchestrated on a pre-reflective level by the kinship of one body with another. Bodies grasp each other thanks to the fact that there is a primordial liaison between oneself and another (Merleau-Ponty 2012). Without having to think about it in abstraction, on an experiential level we grasp moods, modes of conduct, and affective and emotional states in an intuitive sense. As a sensing organ, my body puts me in contact with other bodies not as a recipient of static data, but as a network of constantly unfolding dynamic and expressive meanings. What this means is that notwithstanding the specificity of cultural and affective differences, for the most part social life is regulated by a pre-reflective fluidity that operates on a latent rather than reflective level; such a dynamic is predicated on the idea of the body as an expressive system.

One of the salient aspects of Covid-19 is that it issues a challenge to the phenomenological idea of intercorporeality (cf. Dolezal 2020). This is evident in at least two key respects. First, as expressive and bodily beings, we are always already in touch with other bodies. This is especially true from a Merleau-Pontean perspective. As he sees it, one's own body is not an autonomous mass of materiality solipsistically surveying the world; rather, it is part of a system, which is interwoven in the fabric of other bodies irrespective of our own idiosyncrasies and preferences. Already having a body means being in touch with other bodies, each of whom belongs to the same ontological order of life, and which, thus, form a "single fabric" of being (Merleau-Ponty 2012, 27). This is true as much on a structural level as it is on a sensual level. Just as touch involves a reversible movement between ourselves and the world, so the same is true of other aspects of intercorporeal life, not least *breathing*.

As bodily beings, breathing is not a private practice sealed off from a neutral world; it is a porous—indeed, emblematically *atmospheric*—exchange that reinstates that we are as much in the world as the world is in us. Breathing brings to light in a quite literal way our inheritance with others and our indissoluble liaison in a shared space. True, the manner in which this space is shared (and shareable) is mediated by any number of socio-cultural norms, which either amplify or underplay a sense of space as *ours* rather than *one's own*. Yet from the outset, breathing connects us to a common world, in which our inhalation and exhalation are both biological and affective, as Merleau-Ponty has it, "the body is already a respiratory body. Not only the mouth, but the whole respiratory apparatus gives the child a kind of experience of space" (Merleau-Ponty 1964, 122). It is only later on, when we acquire a sense of breathing as belonging to one's own body that a more rigid boundary line is cultivated between inside and out.

For this reason, breathing is also interwoven with anxiety inasmuch as it indexes a site of vulnerability in our being-in-the-world, as one report attests: "Being around others—especially strangers and crowds—has become an anxiety-ridden proposition. As much as we're yearning to be with people again, we can't help but think of the risks. Is this stranger's cough the one that will infect me?" (Peleg 2020). Thus, just as breathing dissolves the separation of self and other, so it introduces an aspect of anxiety, the manifestation of which is nothing less than breathing itself. Indeed, it is notable that within the history of anxiety, as told from a phenomenological perspective, it is breathing that comes up time and again as the main expression. Here, we can think of Heidegger's account of anxiety as being "so close that it is oppressive and stifles one's breath" or Sartre's account of nausea as a "vision" that leaves one "breathless" (Heidegger 2008; Sartre 1964). In each case, breathing takes shape in the midst of an affective atmosphere, mirroring the surrounding space in terms of being constricted and taut. As the surrounding world becomes oppressive, so our own breathing is felt as a force of oppression, a point that is especially pertinent to Covid-19 insofar as one of the disease's principal symptoms is a shortness of breath.

One of the salient aspects affecting our breathing during Covid-19 is the introduction of face masks. The ubiquity of the face mask is both a marker of a new modality of breathing—now more inwards and self-reflexive—but also a marker of our relations with others. What this discloses is that the face is not insulated by the skin as a protective membrane, nor is the face simply an assemblage of parts; rather, it is a dynamic network, which conveys meaning. Likewise, a mouth is not just a sector of the body employed for consumption and breathing; it is also a space in and through which intersubjective life is given affective expression, as Merleau-Ponty notes: "I perceive the other's grief or anger in his behavior, on his face and in his hands, without any borrowing from an 'inner' experience of suffering" (Merleau-Ponty 2012, 372). As a totality, the eyes do not perform the work of the mouth, as though they were interchangeable and modular. Rather, the face unfolds as a gestalt and when this totality is obscured, the prereflective background upon which human communication takes place—and which is grasped largely in terms of an atmosphere of trust and openness—is broken (cf. Aho 2020). Something else intervenes in this moment, which is often grasped as a mode of suspicion.

Indeed, it is the atmosphere of suspicion that is another key feature of our current relations with others. Stripped of a primary mode of expression, the other has been deprived of their singularity and rendered an anonymous mass of biological (and potentially infected) flesh (cf. Dolezal 2020). Against this, the other's presence is now measured in strictly quantitative terms, underpinned at all times by an anxiety over being too close to strangers, lest they be carriers of the disease even—or especially—unknown to themselves. Indeed, the structure of intercorporeal existence, as it has been instituted in our present era, centres on a series of new practices, each of which demands that we re-habitualise our bodies—often in a counter-intuitive way—to conform to a language of distance and disease. The result is a decisive sense of alienation from both others and to ourselves.

## Conclusion

Let me end by summarising why the concept of atmosphere, as it has been employed broadly within the phenomenological tradition, is singularly beneficial to accounting for the affective structure of Covid-19 in comparison to the terminology of, say, feeling, emotion, or mood. The close connection between the concept of atmosphere and adjoining notions, especially the term "existential feeling" as it has been employed by Matthew Ratcliffe, has meant that generating analytical clarity has often proven difficult in research on atmospheres (Ratcliffe 2008). Nevertheless, while both emotions and feelings tend to index affective structures constituted by individuals—feelings being largely prereflective and tacit while emotions being abstract and sociolinguistic—atmospheres are spatially distributed, not delimited to individuals, and are potentially perceived on an intersubjective and, thus, shared level (cf. Trigg 2020). In this respect, the concept of atmosphere generates a more complex phenomenology than that of emotion alone: atmosphere is both the structure upon which emotions and feelings are instituted while also being a specific kind of affective phenomena which intentionality is itself directed towards.

This double-sided aspect of an atmosphere has been evident in several respects in this chapter. First, I have aimed to demonstrate how the anxiety co-constituted with coronavirus is diffused throughout a given environment in a multi-directional and non-linear way. What this means is that instead of being directed towards a discrete phenomenon, as an atmospheric force, Covid-19 is distributed through the world both on a latent or operative level but also as a thematic and reflective horizon. As much as we talk and reflect about Covid-19 as a particular kind of disease to be treated and managed, so it also already forms a meaningful context from which thoughts and actions emerge. Such is the specific structure of an atmosphere; it is given expression in and through singular things without being reducible to those things, and instead generating a hazy style that permeates everywhere.

And yet as I have also tried to show, an atmosphere has privileged modalities of expression, and in the case of Covid-19, these expressions include home and the body. Where the home is concerned, there is a transformation of the physical home from a sanctuary to a site of restriction, while the surrounding world now appears thematically present as an uncanny terrain, which is pockmarked with risk and danger. The same is no less true of the human body. The body of Covid-19 is not simply an "ill" body that is ravished by disease; it is also an anxious body, a suspicious body, a distanced body, and a concealed body. In a word, the body has become a site of disruption insofar as it indexes a disorder in our normative understanding of the world. Just like home, the body is both foregrounded and simultaneously out-of-joint, both familiar and unfamiliar in the same measure.

In sum, then, the concept of atmosphere can play a critical role as part of a "phenomenological toolkit" in accounting for how complex affective states are distributed in and through a given environment and through multiple subjectivities. As a confluence of subjectivity, materiality, and affect, an atmosphere resists being categorically defined in an analytical way. For that matter, the concept of atmosphere also resists being delimited and reduced to human emotion alone. An atmosphere of anxiety, as it has been dealt with in this study, is as much embedded in spatial configurations—in places and buildings—as it is the crowds and individuals who inhabit those buildings. In this respect, by focusing on the externalisation and materialisation of emotion, atmosphere offers a potentially vital counterpoint to theories of emotion that privilege interior and individual existence.

## References

Aho, Kevin. 2020. "The uncanny in the time of pandemics: Heideggerian reflections on the coronavirus." *The Heidegger Circle Annual* 10, 1–19.

Bergo, Bettina. 2020. *Anxiety: A Philosophical History*. Oxford: Oxford University Press.

Böhme, Gernot. 1993. "Atmosphere as the fundamental concept of a new aesthetics." *Thesis Eleven* 36, 113–126.

Böhme, Gernot. 2017. *Atmospheric Architectures: The Aesthetics of Felt Space*. London: Bloomsbury.

Chesler, Caren. 2020. "Oh, no! Do I have a fever? When coronavirus fears rev up my hypochondria, my 9-year-old keeps me grounded." *Washington Post*, May 16, 2020. https://www.washingtonpost.com/health/oh-no-do-i-have-a-fever-when-coronavirus-fears-rev-up-my-hypochondria-my-9-year-old-keeps-me-grounded/2020/05/15/47285304-8a1d-11ea-ac8a-fe9b8088e101_story.html

Dolezal, Luna. 2020. "Intercorporeality and social distancing." *The Philosopher*. https://www.thephilosopher1923.org/essay-dolezal

Ducharme, Jamie. 2020. "How to soothe your 're-entry anxiety' as COVID-19 lockdowns lift." *Time Magazine*, June 11, 2020. https://time.com/5850143/covid-19-re-entry-anxiety/

Elliott, Larry. 2020. "Half of British adults 'felt anxious about Covid-19 lockdown.'" *Guardian*, May 4, 2020. https://www.theguardian.com/society/2020/may/04/half-of-british-adults-felt-anxious-about-covid-19-lockdown

Griffero, Tonino. 2014. *Atmospheres: Aesthetics of Emotional Spaces*. Trans. Sarah De Sanctis. New York: Routledge.

Griffero, Tonino. 2017. *Quasi-Things: the Paradigm of Atmospheres*. Trans. Sarah De Sanctis. New York: State University of New York Press.

Heidegger, Martin. 1977. *Basic Writings*. Trans. Albert Hofstadter. New York: Harper & Row.

Heidegger, Martin. 2008. *Being and Time*. Trans. John Macquarrie & Edward Robinson. New York: Harper Collins.

Husserl, Edmund. 2001. *Analyses Concerning Passive and Active Synthesis: Lectures on Transcendental Logic*. Trans. Anthony J. Steinbock. Dordrecht: Kluwer Academic Publishers.

Kluger, Jeffer. 2020. "The Coronavirus Pandemic May Be Causing an Anxiety Pandemic." *Time Magazine*, March 26, 2020. https://time.com/5808278/coronavirus-anxiety/

Kurth, Charlie. 2020. "The anxiety you're feeling over covid-19 can be a good thing." *The Washington Post*, April 16, 2020. https://www.washingtonpost.com/outlook/2020/04/16/anxiety-youre-feeling-over-covid-19-can-be-good-thing/

McKie, Robin. 2020. "'I'm broken': how anxiety and stress hit millions in UK Covid-19 lockdown." *Guardian*, June 21, 2020. https://www.theguardian.com/global/2020/jun/21/im-broken-how-anxiety-and-stress-hit-millions-in-uk-covid-19-lockdown

Merleau-Ponty, Maurice. 1964. *The Primacy of Perception*. Trans. William Cobb. Evanston: Northwestern University Press.

Merleau-Ponty, Maurice. 1968. *The Visible and the Invisible*. Trans. Alphonso Lingis. Evanston: Northwestern University Press.

Merleau-Ponty, Maurice. 2012. *Phenomenology of Perception*. Trans. Donald Landes. London: Routledge.

Overy, Richard. 2020. "Why the cruel myth of the 'blitz spirit' is no model for how to fight coronavirus." *Guardian*, March 19, 2020. https://www.theguardian.com/commentisfree/2020/mar/19/myth-blitz-spirit-model-coronavirus

Peleg, Oren. 2020. "Feeling anxious about returning to life in public? You're not alone." *Los Angeles Magazine*, May 4, 2020. https://www.lamag.com/citythinkblog/stay-at-home-lifted-anxiety-germaphobia/

Ratcliffe, Matthew. 2008. *Feelings of Being*. Oxford: Oxford University Press.

Ratcliffe, Matthew, and James Kidd, Ian. 2020. "Welcome to Covidworld." *The Critic*, November 2020. https://thecritic.co.uk/issues/november-2020/welcome-to-covidworld/

Riberio, Celina. 2020. "Return anxiety: 'Coronavirus has caused a mass emotional event in our lives.'" *Guardian*, May 11, 2020. https://www.theguardian.com/world/2020/may/11/return-anxiety-coronavirus-has-caused-a-mass-emotional-event-in-our-lives

Sartre, Jean-Paul. 1964. *Nausea*. Trans. Lloyd Alexander. New York: New Directions.

Schmitz, Hermann. 2011. "Emotions outside of the box—the new phenomenology of feeling and corporeality." *Phenomenology and Cognitive Science*. 10:2, 241–259.

Sumartojo, Shanti, and Pink, Sarah. 2019. *Atmospheres and the Experiential World: Theory and Methods*. London: Routledge.

Trigg, Dylan. 2012. *The Memory of Place: A Phenomenology of the Uncanny*. Athens: Ohio University Press.

Trigg, Dylan. 2016. *Topophobia: A Phenomenology of Anxiety*. London: Bloomsbury.

Trigg, Dylan. 2019. "At The Limit of One's Own Body." *Metodo: International Studies in Phenomenology and Philosophy*. 7:1. DOI: 10.19079/metodo.7.1.75

Trigg, Dylan. 2020. "The role of shared emotion in atmosphere." *Emotion, Space and Society*, May 2020. https://doi.org/10.1016/j.emospa.2020.100658

Wilson, Bee. 2020. "Off our trolleys: what stockpiling in the coronavirus crisis reveals about us." *Guardian*, April 3, 2020. https://www.theguardian.com/news/2020/apr/03/off-our-trolleys-what-stockpiling-in-the-coronavirus-crisis-reveals-about-us

Yuen, K. F., et al., 2020. "The psychological causes of panic buying following a health crisis." *International Journal of Environmental Research and Public Health*. 17:10, 3513. https://doi.org/10.3390/ijerph17103513

# 5 Feeling bodies

## Atmospheric intercorporeality and its disruptions in the case of schizophrenia

*Valeria Bizzari and Veronica Iubei*

### Introducing atmospheres

All of us need air to breathe, not simply to live but to survive. Think how many times, in crucial situations of our life, the air immediately becomes a significant element of our emotional state. We often feel "short of air" when we experience anxiety, distress, or bewilderment; we feel a heavy air when we find ourselves immersed in a collective situation of tension or discomfort; or, we always refer to the air when we breathe deeply and immediately feel calmer, more in control of ourselves. In reality, the above expressions refer to a phenomenon that involves us equally in the pathic-bodily aspect of personhood as in the aesthetic-relational one: namely, the *atmosphere*.

Before going into the history of this concept, thus introducing the most compelling theories produced on the subject so far, we would like to restate that our aim, through this chapter, is to provide new food for thought in today's philosophical debate. Also, in light of recent events sparking global interest, philosophical *atmospherology* may present a possibility for a substantial contribution both to the study of human emotions and to psychopathology (Francesetti & Griffero 2019).

The pre-verbal dimension of the encounter with the outside world acquires its own specific ontological value regarding the atmosphere. Think of entering a room after a bitter argument, during a funeral, or participating in a conference. These are all collective situations that produce a particular impression on us, which sometimes imparts a moulding emotional charge. Atmospheres are powerful phenomena of spatial intercorporeality, endowed with an affective, diffused, and pervasive character. Halfway between feelings and dimensionless spaces, dealing with atmospheres requires us to moderately suspend the usual categories of thought: categories that set phenomena in conceptual and concrete boundaries that are well defined and precise.

From a Neo-phenomenological perspective, the world, with its immanent moods, determines our affective situatedness, in opposition to our *Stimmung* that projects affective tones to the external world. In other words, we should

DOI: 10.4324/9781003131298-6

restart from the body's receptivity (resonance) as the primordial matrix of inter-subjectivity, recognising the phenomenon of atmosphere as a spatial, expansive nucleus of "inter-tunement." "I am already in this world when I say 'I'" (Waldenfels 2004). Before the Cartesian cogito, there are unknown aisles of the body that escape logical-rational apprehension to which phenomenology reserves its passionate investigation, not without sceptic counterparts of cognitivist nature.

Atmospherology might actually be a science of situations, of inter-affective constellations created in collective emotions. Life itself, in its affective articulations, is a "situational living" (Schmitz, Müllan, Slaby, 2011), where the self's corporeity is experienced through its spatial resonance. Atmospheres are not a property of the objects we encounter in everyday life, but modes of being of their occurrence through a space of lived existence.

Nevertheless, what interests Schmitz is to reformulate the concept of phenomenon as a "state of things," rather than a "thing in itself," which is why he revives the phenomenological proposal starting from *pathos*: what reaches us from outside, capturing, impressing, absorbing us in what truly animates interpersonal experiences (Rosa 2019). Healthy atmospheric relationships with other human beings are the only possible means of expression and communication with the lifeworld. In this respect, the existence of the atmosphere induces the visualisation of the situation as a "relational field," not reducible to the individual or the sum of the individuals involved (Francesetti 2014).

From this perspective, also awakening cues from Gestalt therapy (Perls, Hefferline, Goodman 1951), psychopathology can also extend to the study of the so-called "psychopathological fields," based precisely on inter-bodily resonances, atmospheric impressions, and collective pathos. In the first part of the chapter, we will argue for a link between the atmospheric dimension and the arising of collective emotions. In the second part, we will describe a clinical case study where the person suffers from a disruption already at the atmospheric stage, which seems to be impaired and hinders a "sense-making attunement" with others—dramatically detaching the subject from the surrounding world.

## The atmospheric dimension of sociality

The essence of a human being is represented by her corporeality. Our body is the mirror of our life, a living portrait of our vicissitudes. Self-awareness means, first of all, being aware of living in a certain environment and the feeling one has while in it. Beyond the more specific questions concerning the natural, historical, and cultural existence of atmospheres, we propose creating a theory that actualises the dimension of the "in-between" of that pathic encounter between people and spaces that relies on a property of the body: the resonance with the surrounding environment, being-in-tune with it.

Nevertheless, contrasts do exist. The antithetical encounters between environments and human beings are actual events, and they only suggest the existence of a "quasi-thingly" category of collective pathic feelings. However, one must resist a total reification of feelings: as Schmitz writes "they are not invisible clouds, constantly floating in space, continuing to remain in the place where first the gaze turns, and then it turns away" (Schmitz et al. 2011, 105). Atmospheres are different from things because they can be interrupted, come and go, extinguish themselves, and recreate. During a lecture, a serious, professional atmosphere of composed elegance presides. As the lecture ends and the audience moves to leave, this atmosphere dissolves.

According to Schmitz, it is not feelings that are private; it is the emotional involvement that is private. Feeling one's own emotional rapture presupposes a self-attributive act, which is completely subjective. Feelings can be both individual and collective. Think of the waves of anxiety, bewilderment, fear, and confusion that the Covid-19 pandemic solicited from the world. The lived body naturally predisposes itself, through its aisles, to the inter-affective resonance with the atmosphere. Sharing the spatial matrix, the lived body is a "sounding-board" of emotional tones (110). The body aisles correspond through narrowness to oppressive atmospheres and vastness to brightening ones (Griffero 2016). When we feel good, there is a pre-reflective atmospheric correspondence that makes our body so healthy as to be fully immersed in the world and its course. The harmonious chiasm between the atmospheric environment and the lived body is the key to existential well-being.

In the discourse on atmospheres, Merleau-Ponty's intercorporeality would be described as an essential space of resonances, communication, habitus, rhythms; all of those movements that reflect an encounter between people, things, environments. "Communication or the understanding of gestures is achieved through the reciprocity between my intentions and the other person's gestures, and between my gestures and the intentions which can be read in the other person's behavior. Everything happens as if the other person's intention inhabited my body, or as if my intentions inhabited his body" (Merleau-Ponty 1962, 190–195). It is a matter of enhancing that intercorporeal space, in which the atmospheres flutter as authentic quasi-things, ineffable in themselves, sometimes condensed or referred to some point of anchorage, and in any case, become so "external" as to suddenly attack us, take possession of us, and then suddenly disappear.

On the one hand, the body is not only something that insulates us from an undifferentiated totality. The body is also a pathic dimension that resonates with the outside environment, allowing processes of sharing, communication, and osmosis. On the other hand, the body maintains that essential substratum that we feel belonging to us until the smallest cells, which opens up to relationships, allow the self-individuation and the exchange with the external environment. If we then allow our gaze to distance itself from the logical-formal constraints of reflexive consciousness, to capture fluidly the experiential background against which the qualitative aspects of being emerge, it would lead not only to a general reconsideration of *aesthesis* as a source of knowledge, but also to a reformulation of foundational concepts such as "person," "body," and "environment." Therefore, the dimension presently under investigation emotionally involves oneself in the concreteness of sense—where one feels one's body touched by things—how the heartbeat, the respiratory rhythm, and the body temperature vary when coming into contact with the world.

According to Böhme, the question of atmospheres is not confined to the mere acknowledgement of the existence of phenomena with vague contours, yet powerful creators of emotional involvements (Böhme 1998). Living in and with atmospheres is far from being collateral to the phenomenological investigation of the embodied subject and the lifeworld. The way of feeling one's own body, typical of the modern age, where nature is increasingly objectified and artificially manipulated by techno-sciences, is expressed emblematically.

The atmosphere, in Böhme's work, places the perceiver and the perceived on a common ground of synthetic sharing (Böhme 2017). Atmosphere is described more as a condition at the border between the intra- and inter-personal field, an affective situationality, a spatial affordance (Gibson 1966). These conditions, although clearly perceived, are unavailable to linguistically formulated descriptions that could capture their significance: changeable states that have a strong effect on the organism and that decisively orient its action, influence its dispositions and ways of establishing environmental coherence, or endorse certain attitudes and behaviours. In the atmosphere, the body becomes a plasmatic lived-experience, absorbing the pathic features of the space it inhabits; at the same time, it concretises the resonance with the atmosphere by modulating its own mode of feeling.

## A taxonomy of sociality: from atmosphere to shared emotions

We can observe our embodied engagement with the world in different levels of interaction. Being embodied not only accommodates self-awareness but also cultivates an understanding of others.[1] In other words, representational knowledge is not responsible for our intersubjective perception. Rather, our affectivity seems to result from an implicit knowing, as Fuchs writes, "an embodied, intuitive knowledge of how to interact with others, how to have fun together, how to elicit attention, to avoid rejection etc. It is a temporally organised, 'musical' memory for the rhythm, dynamics and undertones which resonate in the interaction with others" (Fuchs 2016a, 223). Critically, this "embodied knowledge" arises *before* higher cognitive functions.[2] Merleau-Ponty was the first to describe *intercorporeality* as a process based on the immediate transfer of corporeal schema. In this view, intercorporeity is a pre-reflective and "lived in" type of knowledge that allows the subject to recognise the other in an immediate and non-thetic manner.

> The reason why I have evidence of the other man's being-there when I shake his hand is that his hand is substituted for my left hand, and my body annexes the body of another person in that "sort of reflection" it is paradoxically the seat of. My two hands "coexist" or are "compresent" because they are only one single body's hand. The other person appears through an extension of that compresence; he and I are like organs of one single intercorporeality.
>
> (Merleau-Ponty 1964, 168)[3]

The body appears to be the place where (shared) meaning emerges. In the case of an intersubjective encounter, intercorporeality is that pre-reflective intertwining of lived and living bodies that mutually resonate with one another without requiring inferential capacities. It is that mutual bodily synchrony that allows two subjects to experience subjective and objective qualities through their lived bodies. Furthermore, the other's bodily consti-tution can also be considered the first stage of the constitution of the shared objective world. At this level, there exists a sense of belonging to a com-munity or a group that transcends the individual while it permeates every moment of her life.

The highest levels of intersubjective attunements are collective emotions,[4] which include group-based and shared emotions. On the one hand, the sub-ject A feels the emotion X as a member from the group Y. On the other hand, the involved subjects experience the emotions in unity; they feel the emotion together. For group-based emotions, the crucial component seems to be the subjective membership to a specific group: We can claim that for the arising of a shared emotion, the *bodiliness of feelings*, as Thonhauser calls it (2021), seems to be necessary. In our view, the bodiliness of feelings is understood as atmospheric intercorporeality. To synthesise, our corporeality is neither confined to our physical body nor bounded to our individuality. Rather, it transcends both the material and the individual dimensions. This is the main reason why we find ourselves immediately absorbed in interaffective engagement; we manage to synchronise and tune in with the feeling of oth-ers, to experience the same feeling like a "we" in a "plural pre-reflective and non-thematic self-awareness" (Schmid 2015, 51).[5]

Central for the development of shared emotions is not merely the experi-ence of oneself as a member of a group but something even prior, something mediated by our lived body that allows the subjects to tune in with the oth-ers and the world. In a similar way, Merleau-Ponty suggests that individu-als adopt affective styles, characteristic bodily manners of behavior (such as recurrent gestures or postures), that are typical of their socio-cultural background (Merleau-Ponty 1962). We can conjecture that affective styles facilitate the arising of shared emotions.

In this view, atmospheric intercorporeality seems to be a central feature of shared emotions, something necessary for the arising of a collective attunement.[6] If we follow this characterisation, we can indeed argue that a disruption at an intercorporeal atmospheric level will cause impairments in all of the other layers of intersubjective engagement, through an avalanche effect, which can be more or less severe according to the pathology. One clinical study that we view as paradigmatic of a deep, intercorporeal disruption already present at the atmospheric stage is schizophrenia. In such cases, the atmosphere is peculiar even before the arising of the psychotic experience. For this reason, we aim at enriching the already existing literature by linking the disruption of the *atmospheric intercorporeality* to that of higher levels of social engagement.

## Schizophrenia: the collapse of a shared world

Psychopathologies often involve the disappearance of the lived body as the centre of the being: suddenly, it becomes reified and alien. In the dialectical relationship that usually links the body-that-I-am and the body-that-I-have, we can observe a prevailing stiffening of the body-object, to such an extent that there is an alienating feeling of strangeness towards it. Further, the body is no longer able to recognise its boundaries—the link with the external reality represents a disruptive event. This is what happens in schizophrenia,[7] where the living body (*Leib*) is reified, and the subject can also identify herself with an object through the process of depersonalisation. While the schizoid patient perceives her body as a cumbersome container for the self, the schizophrenic patient endures a fragmentation of the body, and accordingly, a fragmentation of the self-experience. As noted by Fuchs and Röhricht (2017), the process of disembodiment tends to involve the following features: a weakening of the basic sense of self; a disruption of implicit bodily functioning; and, a disconnection from the inter-corporeality with others.

Also, in terms of perception, the ability to recognise familiar patterns of objects could be impaired, and the subject may register a disintegration of habits or automatic practices. In other words, in the schizophrenic subject *praktognosia* (speaking in the Merleau-Pontian sense), our tacit and enactive knowledge is lost. Considering that, "[t]o have a body is to possess a universal setting, a schema of all types of perceptual unfolding and of all those inter-sensory correspondences which lie beyond the segment of the world which we are actually perceiving" (Merleau-Ponty 1962, 326), we can affirm that being schizophrenic means losing these sets of possibilities, or affordances.

The practical immersion of the self in the world, normally mediated by the body, is impaired or lost— hindering the facilitation of active, intersubjective engagements. In other words, this "disembodiment" (cf. Fuchs 2005; Stanghellini 2004; Bizzari 2018) of self implies not only the impairment of self-awareness but also the loss of social attunement, resulting in a fundamental alienation of intersubjectivity (cf. Sass and Pienkos 2013). The basic sense of being-with-others is replaced by a sense of detachment that may pass into a threatening alienation. It is very interesting to notice that the loss of the spontaneous attunement with the world, usually mediated by the body, can appear many years before the onset of acute psychosis.

## The prodromal stages of Schizophrenia

"The environment is somehow different—perception is unaltered in itself but there is some change which envelops everything with a subtle, pervasive and strangely uncertain light" (Jaspers 1997, 82). In his analysis of psychotic disturbances, Jaspers draws attention to a singular experience lived by the patient in the prodromal stage of schizophrenia. Defined as "delusional mood" or "delusional atmosphere" this experiential phase is characterised by a diffused sense of unreality, estrangement, and loss of natural evidence (Blankenburg 2001). The person feels "as if something is going to happen"; something crucial, mysterious, oppressing, sometimes even apocalyptic. We are dealing with a period of life that feels highly anxious to the subject in question: she lives on the verge of experiencing something that will radically change her life, revealing the content of her anxiety and fear. Conrad accurately describes this state in his book covering the early onset of schizophrenia (Conrad 2012), collecting and analysing stories of young men at the edge of their career, who suddenly became extremely worried, anxious, and even terrified by the vague feeling that the general stream of life has changed, perhaps threatening a major, extraordinary event that could dramatically perturb their world. Far from a satisfactory understanding of this phase, Conrad hypothesises that it is the primal (or basal) affectivity, i.e., the atmosphere of affective attunement with the interpersonal space, which is disrupted in the patient at this point.

Even before the collapse into a schizophrenic episode, the person complains about a social impoverishment: her own existential disconnection from the natural, granted background of life. She speaks of feelings of alienation and dread in a space that emanates an "atmosphere of heavy expectation," which progressively assails her lived-body, forcing it to an affective, solipsistic contraction (Schmitz et al. 2011). In this condition, which is not yet a proper delusion but prepares the delusional experience, each kind of communication is corrupted and interrupted. Interpersonal articulations seem to have indeterminate and indefinite connotations. Objects assume modified meanings, falling into a radical inconsistency, into an ontological and semantic fading from which no structured unity of sense can emerge. This "uncanny particularity" (Sass 1994) also involves the way the subject perceives the other—compromising the immediate and spontaneous attunement usually provided by intercorporeality. As a result, immediate attunement is replaced by a feeling of confronting a world that is fragmented, meaningless, unreal, yet often insinuating. In our view, this atmospheric stage is essentially grounded in an affective corporeality: the ontological, bodily openness towards the others and the world, prior to any emotional experiences and knowledge.

Therefore, what begins to emerge at this stage seems to be a disturbance of the atmospheric, interpersonal background, which we can evaluate as a fusion between Schmitz's concept of atmosphere and Blankenburg's concept of "natural evidence" (2001). In Blankenburg's view, schizophrenic patients complain about a loss of natural evidence, which is a latent feeling of familiarity, a sense of comforting trust in the world that permeates everyday life and interpersonal relationships. Tacit evidences collapse dramatically, becoming problematic; the capacity of experiencing falls apart, and one's own ontological security is replaced by a sense of disorienting and unsafe existential conditions. As described by Conrad:

> We should imagine it as the ground-structure of the subject's field, who walks alone in a dark forest; nothing is natural, nothing is evident. In the darkness, they are lurking—without knowing what is lurking. What frightens us is not the tree or the bush we see, nor the owl's screech we hear, but all that constitutes the background ... it deals with the obscurity and the background themselves.
>
> (Conrad 2012, 59, V.I. translation)

There is the loss of spontaneous attunement with the world (common sense), whereby new things become spontaneously incorporated into the subject's view. The familiar space blurs and eventually falls apart. The boundaries between the self and the world are indistinguishable, and the subject can also lose her self-awareness: "Sometimes I feel as if I do not have an ego anymore, and I blend in with the world. I look at a flower and the flower is not anymore distinct from me: the flower and I are the same thing. I am a detached observer who sees the flow of life where everybody runs around. I am aware I am part of this flow but there is no distinction between myself and the objects: everything is chaos" (Borgna 2015, 120, V.B. translation).

Merleau-Ponty seems to grasp this condition when he faces the importance of space as a structure that allows us to inhabit a shared world, something that is lost in schizophrenia: "The schizophrenic no longer inhabits the common property world, but a private world, and no one gets as far as geographical space: he dwells in 'the landscape space,' and the landscape itself, once cut off from the common property world, is considerably impoverished .... everything is amazing, absurd or unreal" (Merleau-Ponty 1962, 287).

Especially in the pre-psychotic stage, this atmospheric disruption is accompanied by a paradoxical mixture of immobility and protention in the experience of time (cf. Stanghellini, Mancini, 2017) as if it was suspended. In this way, temporal fragmentation contributes to the fragmentation of self and to social detachment, leading to particular and uncanny experiences such as premonitions about oneself and the external world. In a similar manner, space is no longer a space of possibilities (or affordances), but it is conceived by the schizophrenic person as an unknown territory whose objects (and persons) are fragmented and threatening.

The break of the atmospheric intercorporeality develops into specific symptoms which are explicitly involved within the intersubjective capacities: hypo-attunement (the diminished emotional contact with other people and, consequently, the inability to understand "others" behaviour), and invasiveness (problems in self-other awareness and feeling of being oppressed by others both in terms of bodily sensations and in terms of emotional overwhelming). These symptoms elicit the subject to adopt compensatory strategies, such as an algorithmic conception of sociality and an antithetic attitude, that is, the decision of the subject to build her own existential values against common sense.[8] This is a response towards this feeling of vulnerability that emerges from a compromised atmospheric intercorporeality that has become an uncanny and unfamiliar threat to the self. It seems that this atmospheric stage is, therefore, characterised by an egocentric, centripetal component, which leads to an abnormal a-specificity, diffusion, and dispersal of the intentional meaning.[9]

## The autobiography of a Schizophrenic girl

A famous case study described by Sechehay (1962) in *Autobiography of a Schizophrenic Girl* is instructive here. In particular, the development of the pathology, as described in the book, is beneficial for understanding how the atmospheric dimension plays a key role in hindering the arising of shared emotions. The book gives a detailed account of a real case study and is a substantial first-person report from an individual who suffered from psychotic experiences before, after a troubled path into the pathology, finally recovered, crediting her therapeutic regime. Here, we won't give a comprehensive account of the story; instead, we will focus on the most important steps through which Renee entered and lived her journey. At the beginning, she describes the "Appearance of the first feelings of unreality":

> I remember very well the day it happened. We were staying in the country, and I had gone for a walk alone as I did now and then. Suddenly, as I was passing the school, I heard a German song; the children were having a singing lesson. I stopped to listen, and at that instant a strange feeling came over me, a feeling hard to analyse but akin to something I was to know too well later—a disturbing sense of unreality ... Everything was exact, smooth, artificial, extremely tense; the chairs and tables seemed models placed here and there. Pupils and teachers were puppets revolving without cause, without objective. I recognized nothing, nobody. It was as though reality, attenuated, had slipped away from all these things and these people.
>
> (Sechehay 1962, 21–26)

The feeling of unreality shapes both the objects and the people, who Renee perceives as puppets and deanimated automata. They are not lived-bodies anymore. Accordingly, it is impossible to establish an emotional bond with them. The loss of the atmospheric intercorporeality is the source of this sense of isolation and the solipsistic delusion that if one desisted from creating everything, the world would cease to exist. This feeling of unreality changes and becomes a real struggle. For example, in the chapter "The struggle against unreality begins," Renee argues: "I sensed again the *atmosphere of unreality*. The unreal situations increased ... unreality had crept over people, even friends. It was a torment" (29).

In face-to-face interactions, Renee experiences persistent problems: "During the visit I tried to establish contact with her, to feel that she was actually there, alive and sensitive. But it was futile. Though I certainly recognized her, she became part of the unreal world. I knew her name and everything about her, yet she appeared strange, unreal, like a statue" (36).[10] The disorder seems to move from the atmospheric stage to the concrete meeting with the other, and it also seems to involve the corporeal dimension, collapsed into the organic, static, and motionless body. It is, therefore, impossible for intercorporeality to emerge. When the disorder becomes chronic, these experiences seem flattered and more detached, yet also involve changes in the atmospheric dimension: often manifesting as a looming sense of meaninglessness, devitalisation, personal irrelevance, and unreality.

If we keep on following Renee's story, we can notice that the central chapter of the book is indeed entitled "I sink into unreality," and that the distance between her and the intersubjective dimension seems to grow bigger and bigger: "Living in an environment empty, artificial and apathetic, an invisible, insuperable wall divided me from people and things. I saw few people and wanted only to be alone, hiding in the cellar where, sitting on the coal pile, I remained quiet, unmoving, my gaze fixed on a spot or a gleam of light. But behind this wall of indifference, suddenly a wave of anxiety would creep over me, the anxiety of unreality" (81).

The alterations in the domain of the lived world are intertwined with alterations in the experience of the lived body: in fact, it is the lived body that conveys the practical knowledge of how to interact with others, how to understand their expressions and actions, contextualised by familiar situations. Thus, we are involved in a sphere of primary intercorporeality, a tacit and enacted knowledge, that is also the basis of "common sense." Primary intercorporeality provides a fluid, automatic, and context-sensitive pre-understanding of everyday situations, connecting self and world through a basic habituality and familiarity. As we argued, this kind of involvement in the world is disturbed in schizophrenia and results in a fundamental alienation of intersubjectivity.

This intersubjective alienation not only affects the so-called "primary intersubjectivity" but also produces higher levels of collective involvement, such as shared emotions. In other words, what is at stake in schizophrenia is not a mere disruption of the "I-Thou" relationship but the ability itself to feeling part of a "we": the sense of being together and feeling together. We argue that this necessary and pre-reflective attunement is atmospheric intercorporeality: central for the development of shared emotions. Therefore, it is not merely the experience of oneself as a member of a group but something which precedes the experience of belonging; something mediated by our lived body which allows subjects to tune-in with others and the world.

In schizophrenia, intercorporeality registers severe impairments, hindering the arising of shared emotions and collective engagements. From our perspective, it seems clear that the core of the disruption lies in something which belongs to the pre-inferential, intercorporeal domain. It is not a coincidence that Renee, in the last chapter on her recovery, claims: "I learn to know my body" (109). Meaning that, little by little, she was re-acquiring her intercorporeal awareness, filling the gaps between herself and other people.[11]

## Conclusion

In this chapter, we analysed schizophrenia through the lens of atmospheres. In particular, we focused on the prodromal stage of this pathology, emphasising the fact that the subject is detached from the social world and characterised by an anomalous or absent embodied being. Being bodily engaged with the other seems a necessary requirement for experiencing higher levels of sociality, such as collective emotions. Although many authors identify the real core of schizophrenia as a mere disruption of the self, we argued that the main important deficit lies in the atmospheric stage—a stage which is not only embodied but also already linked with the social realm.

To support our view, we examined a famous case study. In doing so, we posited two key aims: clinical and philosophical. From a clinical perspective, focusing on the atmospheric stage allows to enhance the diagnostic process and enrich the therapeutic dimension, which begins before the emergence of psychoses (cf. Bollas 2013). From a philosophical point of view, the case study of schizophrenia represents a solid ground to start re-evaluating the atmospheric stage concerning higher orders of social engagement. In fact, the field of application of the notion of the atmosphere is truly vast: ranging from architecture to advertising, from "commodity aesthetics" (of which Böhme offers a varied sample of examples within the consumerist trend), to art, and finally, as we have illustrated through the case of schizophrenia, to phenomenologically-oriented psychiatry. It is clear that, if seriously pursued and distanced from old prejudices surrounding the valueless inferiority of phenomena which are not "scientifically provable," the philosophy of atmosphere could constitute an important contribution to current debates on the phenomenology of psychopathology.

# Notes

1 The fact that we are innately intersubjective and open to the others, that are implied in our phenomenal field since the beginning, has been argued by Husserl himself. Here is a central passage from the *Crisis*: "(…) within the vitally flowing intentionality (*ebendig strömende Intentionalität*) in which the life of an ego-subject consists, every other ego is already intentionally implied from the very start by way of empathy and the empathy-horizon (*Einfühlungshorizont*). Within the universal epoché (…), it becomes evident that there is no separation of mutual externality at all for souls in their own essential nature. What is a mutual separateness (*Außeinander*) for the natural-mundane attitude of world-life prior to the epoché, because of the localization of souls in living bodies, is transformed in the epoché into a pure, intentional interweaving (*wechselseitiges intentionales Ineinander*)" (Hua 1954, 259).

2 This is testified by studies in developmental psychology (see, for example, Stern 1985), but also by the analysis of the autism spectrum disorder, where the intersubjective deficit seems to derive from a disturbance of embodied interaction and not from a merely cognitive disruption (Fuchs 2015, 225).

3 Flesh as correlation is intended as a system of articulation of different beings and things; these elements share the same thickness of the sensible in virtue of their bodily dimension. In using the expression "organs of one single intercorporeality," Merleau-Ponty is expressing the idea that this co-presence is literally carnal. The intercorporeality that characterises this experiential situation does not imply a mere presence of different discrete entities, but rather an articulation of entities that share a general carnality. At this stage, we know that the flesh has a character of generality, all organic and inorganic objects and beings have a carnal presence. The idea of the flesh is the starting point for the phenomenological reflection, this carnality—in its intersubjective dimension—is what allows the constitution of the world and beings (human or animal). See also, Bizzari & Guareschi 2017, Guareschi 2019.

4 For an overview of the debate on collective emotions see von Scheve and Salmela (2014).

5 In accounting for group-directed empathy, Salice and Taipale (2015) argue that there might be a *collective body* with an expressiveness that is not reducible to individual bodies.

6 Following the work of Szanto (2016; 2018), we can also add as necessary for shared emotions the *plurality requirement*, i.e., awareness of the plurality of partaking individuals; and the *integration requirement*, i.e., a sense of togetherness characterising the experience.

7 The description of schizophrenia that we can find in the Diagnostic and Statistical Manual of Mental Disorders (DSM), Fifth Edition, is the following: "The essential features of schizophrenia are a mixture of characteristic (…) positive and negative [symptoms] that have been present for a significant portion of time (…), associated with marked social and occupational dysfunction" (APA, 2013).

8 The erosion of the atmosphere seems to have a semantic correlation: in schizophrenia, reality assumes another meaning, another depth, and the subject uses alternative symbolic meanings. As noted by Hemmo Müller-Suur (1962), there are different levels in which the symbolic experience changes. He describes this process in terms of a symbolic deconstruction of reality. The first modification appears when the subject includes meanings that come from a mythological background (this is observable in patients' draws which represent legends, myths or fairy tales). The second change happens with the destroy of symbolic meanings, which become disembodied and dissolving. Reality is

locked in abstracts and empty pictures. There is a tendency to reify meanings and reality, to cut off emotions in favour of a formal abstraction. This process culminates with a "nullification" of content. We can claim that the collective dimension is disrupted also at a semantic level.

9   It is important to recall Bruno Callieri's theory (1962) according to which the atmospheric stage of schizophrenia is not a mere prodrome to the psychosis, but owns its distinct ontological reality, it is an independent and autonomous lived experience.

10  It is interesting to notice that it is quite common to find people with schizophrenia who describe others as "statues." In a beautiful text, Hugo von Hoffmannnsthal (1902) seems to grasp the intercorporeal break that is at stake in the pre-psychotic stage: "I fell into a feeling of a threatening loneliness; I felt as if I were trapped in a garden, surrounded by eyeless statues.." (VB's translation). Here the emphasis is, in fact, not only on the sense of detachment from the common world, but also on the others, who are perceived as *Koerper* and not as *Leiber*, as eyeless statues with whom is impossible to communicate. Intercorporeality is impossible to take place, and the subject experiences this sense of being locked in the individual dimension. The living space as an ontological and epistemological category, as an essential structure of the ego, is disrupted and deprived of those features which usually allow to immediately perceive the world as familiar.

11  Another first person reports from Borgna (2015) seems to confirm our hypothesis: "I feel good, now I see the trees, the water, the reality, and I am quieter. My material body walks freely and I give more value to friendship, to stay together and not to money or other things" (Borgna, 2015, 120). Here we can notice the link between the "functioning" body and the collective emotions, such as friendship.

## References

American Psychiatric Association (APA). 2013. *Diagnostic and Statistical Manual of Mental Disorders: DSM 5*, 5th ed., Arlington (VA): American Psychiatric Association.

Bizzari, Valeria., Guareschi, Carlo. 2017. "Bodily Memory and Joint Action in Music Practice and Therapy." In: *Quaderni della Ginestra, special issue on Philosophy and Collectivity*, 3/2017, 114–121

Bizzari, Valeria. 2018. "Schizophrenia and Common Sense: A Phenomenological Perspective." In *Schizophrenia and Common Sense: Explaining the Link between Madness and Social Values*, ed. by Hipolito et al. Berlin: Springer, Vol. 12, 39–53.

Blankenburg W, Mishara AL. 2001. First steps toward a psychopathology of "common sense". *Philosophy, Psychiatry, & Psychology* 8(4), 303–315.

Böhme, Gernot. 1998. *Anmutungen. Über das Atmosphärische*. Stuttgart: Edition Tertium.

Böhme, Gernot. 2017. *The Aesthetics of Atmospheres*. New York: Routledge.

Bollas, Christian. 2013. *Catch Them Before They Fall. The Psychoanalysis of Breakdown*. London: Routledge.

Borgna, Eugenio. 2015. *Come se Finisse il Mondo. Il Senso Dell'Esperienza Schizofrenica*. Milano: Feltrinelli.

Bruno, Callieri. 1962. "Aspetti psicopatologico-clinici della "Wahnstimmung"." In Kranz H (ed). *Psychopathologie Heute*. Stuttgart: Thieme.

Conrad, Klaus. 2012. *La Schizofrenia Incipiente. Un Saggio di Analisi Gestaltica Del Delirio*. Rome: Giovanni Fioriti Editore.

Francesetti, Gianni, 2014. "Dalla Sintomatologia Individuale ai campi psicopatologici. Verso una prospettiva di campo sulla sofferenza clinica." *Quaderni di Gestalt*, XXVII 2.

Francesetti, Gianni, Griffero, Tonino. 2019. *Psychopathology and Atmospheres.* Newcastle upon Tyne: Cambridge Scholars Publishing.

Fuchs. Thomas. 2005. "Corporealized and disembodied minds: a phenomenological view of the body in melancholia and schizophrenia." *Philosophy, Psychiatry and Psychology* 12, 95–107.

Fuchs, Thomas. 2015. "Pathologies of intersubjectivity in autism and schizophrenia." *Journal of Consciousness Studies*, 22(1–2): 191–214.

Fuchs, Thomas. 2016a. "Embodied Knowledge, Embodied Memory." In Rinofner-Kreidl, Sonja, Wiltsche, Harald. (Eds.) *Analytic and Continental Philosophy. Methods and Perspectives. Proceedings of the 37th International Wittgenstein Symposium.* Berlin: De Gruyter, 215–229.

Fuchs, T., Röhricht, F. 2017. "Schizophrenia and intersubjectivity: An embodied and enactive approach to psychopathology and psychotherapy." *Philosophy, Psychiatry, and Psychology* 24 (2), 127–142.

Gibson, J. J. 1966. *The Senses Considered as Perceptual Systems.* London: Allen and Unwin.

Griffero, Tonino. 2016. *Il Pensiero dei Sensi: Atmosfere.* Milano: Guerini & Associati.

Guareschi, Carlo. 2019. Merleau-Ponty's philosphy of nature and intercorporeality: *An embodied model for contemporary environmental aesthetics.* PhD Thesis, University College Cork.

Husserl, Edmund. 1954 (Hua VI). *Die Krisis der Europäischen Wissenschaften und Die transzendentale Phänomenologie. Eine Einleitung in Die Phänomenologische Philosophie.* Ed. by Walter Biemel. Den Haag: Nijhoff.

Jaspers, Karl. 1997. *General Psychopathology* (7th ed.). Trans. J. Hoenig, M.W. Hamilton, Baltimore (MD): Johns Hopkins University Press.

Merleau-Ponty, Maurice. 1962. *Phenomenology of Perception.* London: Routledge.

Merleau-Ponty, Maurice. 1964. *Signs.* Evanston: Northwestern University Press.

Müller-Suur, Hermo. 1962. *Das Schizophrenie als Ereignis. Psychopathologie Heute.* Stuttgard: Thieme.

Perls, Frederick, Hefferline, Ralph F., Goodman, Paul. 1951. *Gestalt Therapy.* New York: Julian Press.

Rosa, Hartmut. 2019. *Resonance: A Sociology of Our Relationship to the World.* Oxford: Polity Press.

Salice, Alessandro, Taipale, Joona. 2015. "Group-directed empathy: A phenomenological account." *Journal of Phenomenological Psychology* 45, 163–84.

Sass, Louis. 1994. *The Paradoxes of Delusion: Wittgenstein, Schreber and the Schizophrenic Mind.* New York: Cornell University Press.

Sass, Louis, Pienkos, Elisabeth. 2013. "Space, time and atmosphere. A comparative phenomenology of melancholia, mania and schizophrenia." *Journal of Consciousness Studies*, 20(7–8), 131–52.

Schmid, Hans Bernard. 2015. "On Knowing What We're Doing Together: Groundless Group Self-Knowledge and Plural Self-Blindness." In Brady, Michael, Fricker, Miranda (eds.) *The Epistemic Life of Groups. Essays in the Epistemology of the Collectives.* Oxford: Oxford University Press.

Schmitz, Hermann, Müllan, R.O., Slaby, J. 2011, Emotions outside the box- the new phenomenology of feeling and corporeality. *Phenomenology and Cognitive Sciences* 10, 241–259.

Sechehaye, Marguerite. 1962. *Autobiography of a Schizophrenic Girl*. New York: Penguin.

Stanghellini, Giovanni. 2004. *Disembodied Spirit and Deanimated Bodies*. Oxford: Oxford University Press.

Stanghellini, Giovanni, Mancini, Milena. 2017. "The Life-World of Persons with Schizophrenia." In *The Therapeutic Interview in Mental Health: A Values-Based and Person-Centered Approach*. Cambridge: Cambridge University Press.

Stern, Daniel. 1985. *The Interpersonal World of the Infant: A View from Psychoanalysis and Developmental Psychology*. New York: Basic Books.

Szanto, Thomas. 2016. "Husserl on Collective Intentionality." In A. Salice and H. B. Schmid (eds.) *Social Reality: The Phenomenological Approach to Social Reality. History, Concepts, Problems*. Berlin: Springer, 145–72.

Szanto, Thomas. 2018. "The Phenomenology of Shared Emotions: Reassessing Gerda Walther." In Luft, Sebastian & Hagengruber, Ruth (eds.), *Woman Phenomenologists on Social Ontology*. Cham: Springer, 88–104.

Thonhauser, Gerhard. 2021. "Shared Emotions and the Body." *Danish Yearbook of Philosophy* (published online ahead of print 2021). https://doi.org/10.1163/24689300-bja10004

von Hoffmannnsthal, Hugo. 1902. *Ein Brief*. In: Der Tag. Berlin.

von Scheve, Christian, Salmela, Mikko. 2014. *Collective Emotions: Perspectives from Psychology, Philosophy, and Sociology*. Oxford: Oxford University Press.

Waldenfels, Bernard. 2004. "Bodily experience between selfhood and otherness." *Phenomenology and the Cognitive Sciences*, 3, 235–248.

# 6 Agency and atmospheres of inclusion and exclusion

*Joel Krueger*

## Introduction

Atmospheres seem to be everywhere. They are a central part of everyday life. We make decisions about who to spend time with, what to put on our walls, and where to eat based on atmospheres we associate with people, things, and spaces. In this way, atmospheres shape our experience and behaviour. However, the link between atmospheres and agency has been relatively underexplored in the philosophical literature. Much of the debate instead concerns the *nature* of atmospheres, what sort of things they *are*.

This focus on the ontology of atmospheres is a rich and philosophically substantive area of work. However, in what follows, I argue that it potentially overlooks important insights into the *regulative* power of atmospheres, that is, their capacity to shape the things we do and the ways we connect (or fail to connect) with others. Atmospheres do things. They actively shape experience and behaviour—and crucially, they open up (or close down) forms of social connectedness. They do these things, I argue further, because atmospheres don't merely provide affective colour or texture. They also furnish *possibilities*—possibilities that help or hinder us as we find our way in the world. I unpack this claim by considering atmospheres as "affective arrangements" (Slaby, Mühlhoff, and Wüschner 2017). Along the way, I develop a distinction between "atmospheres of inclusion" and "atmospheres of exclusion," and I apply this distinction to two case studies: Sara Ahmed's critical phenomenology of "stopped bodies," and social difficulties in autism. Both of these cases, I conclude, help to highlight the deep connection between atmospheres and agency.

## Preliminary remarks

We often speak of atmospheres as though they pick out some well-defined entity—a quality, feature, attribute, or presence that attaches to the people, things, and spaces we encounter in everyday life. We talk about the tranquil atmosphere of a spring morning or lush park. A child can radiate an atmosphere of boundless curiosity, hope, innocence, and enthusiasm.

DOI: 10.4324/9781003131298-7

The family dog can seem trusting and earnest or skittish and twitchy. A piece of music might emanate a sombre or uplifting atmosphere. A classroom can feel lively and inclusive, or threatening and closed-off to open inquiry. Some may experience a family dinner as authentic, loving, spontaneous, and warm; for others, it might feel grim and stilted. Homes, workplaces, churches and temples, restaurants, heritage sites, sports venues, stores, clubs, museums, theatres, factories, music venues, parks, cities, persons, activities, and communities—among many other things—are all said to have distinctive atmospheres.

When we speak of atmospheres in everyday life, we generally don't simply take an aesthetic interest in them the way, say, we might view a work of art (although we can, of course, consider atmospheres from a purely aesthetic point of view). Our interest tends to be more concrete. This is because atmospheres aren't causally inert. They pervade everyday life in ways that shape our evaluations and behaviour. In this way, they can be said to have *practical* significance.

For example, we might avoid courses taught by a prickly professor who, despite their technical brilliance and international prestige, creates a notoriously unpleasant classroom atmosphere. We may skip a restaurant with well-prepared food because we don't like the colourless or out-of-date atmosphere (although such an atmosphere may, for some, be an attractive part of its quirky or kitschy charm), or routinely work in a coffeehouse with middling coffee because we like the cosy atmosphere and find it conducive to working. The key point—important for what follows—is that atmospheres *do things*. They envelope us, press upon us, and in so doing play an important role in shaping how we evaluate and get on in the world. They help or hinder as we "find our way," as Sara Ahmed (2006) puts it.

Yet, despite the ease with which we speak of atmospheres and their presumed ubiquity, there is, nevertheless, surprisingly little consensus about what atmospheres *are*. Atmospheres are now increasingly discussed in a variety of disciplines and debates, from architectural studies, aesthetics, and management studies to psychology, geography, anthropology, and sociology (see Osler and Szanto (this volume) for a helpful overview). Discussions are also on the rise in philosophy, as the present volume attests. However, despite this increased attention, there is still widespread recognition that atmospheres are "slippery" phenomena (Böhme 1993), difficult to pin down with ontological precision. Part of this slipperiness comes from the fact that, as we've already noted, atmospheres traverse distinctions between people, things, and spaces (Anderson 2009). It's possible that most things could be described as atmospheric.

The way we talk about atmospheres in both academic discourse and everyday life highlights this ontological ambiguity. On the one hand, atmospheres are often spoken of as though they have a mind-independent reality. They are said to be out in the world, features of the natural and built environment that "seem to fill the space with a certain tone of feeling like a haze" (Böhme 1993, 113–114). This way of speaking has led some to characterise atmospheres as "quasi-objective" (Böhme 2006) or "quasi-things" (Griffero 2014). Similarly, Schmitz (2019) characterises atmospheres as a kind of pre-personal affectivity that circulates through public spaces. However, there is, on the other hand, also recognition that atmospheres are experiential phenomena; they are *felt*. It seems puzzling to think of an atmosphere as existing somehow out in the world without the presence of a subject (or subjects) who feel it. Additionally, the descriptive richness and specificity of atmospheric qualities we ascribe to people, things, and spaces—e.g., serene, homely, strange, stimulating, holy, melancholic, uplifting, depressing, pleasant, moving, inviting, erotic, collegial, open, sublime, etc. (Böhme 1993)—highlights the fact that atmospheres are, in some important sense, *modes of encounter* (Osler 2021). They are ways that, as embodied subjects, we experience our world, as well as our relation to that world and the people in it. In this way, atmospheres seem to sit uneasily between purely subjective or objective characterisations (Slaby 2019).

I have neither the interest nor the ability to sort out this ambiguity in what follows. Again, this is because my focus is not so much on what atmospheres are but rather what they do. I am specifically concerned with the interrelation between atmospheres and *agency*. The latter, as I use the term here, encompasses both action and affect. Atmospheres do things by animating and regulating actions at both individual and collective levels. They also animate and regulate affective experiences—again, at both individual and collective levels. Atmospheres do this regulative work by opening up—and also closing down—possibilities for emotional experience, behaviour, and social connection.

In light of my focus on agency, I will adopt a *relational* conception of atmospheres. Atmospheres, as I think of them, are tied to the world; they are rooted in features of the natural and built environment. These natural, structural, and organisational features are what make certain kinds of atmospheric experiences—including the emotional expression and behaviour that are part of these experiences—possible. They are what give bits of the world a distinctive "tone of feeling." However, these features are, on their own, insufficient to generate atmospheres. Bits of the world may be configured in ways that are poised to potentially generate atmospheric experiences. But atmospheres only arise if subjects are also present and poised to potentially *engage* with them in some way. Atmospheres, as I speak of them, are intrinsically experiential. Again, it is telling that many of our descriptions involve phenomenal concepts. We try to get a grip on the vagueness and ontological ambiguity of atmospheres by tracing them back to how we experience and live through them. When describing atmospheres in everyday life, we tend to describe both features of the world as well as our felt connection to those features.

## Atmospheres as affective arrangements

With that background in place, I now turn to a more focused considera-tion of atmospheres and how they relate to agency. I will try to side-step some of the ontological puzzles mentioned above by thinking about atmos-pheres in the context of recent interdisciplinary approaches to situated cognition—and more precisely, *situated affectivity* (e.g., Colombetti and Krueger 2015; Colombetti and Roberts 2015; Griffiths and Scarantino 2009; Krueger 2014b; Saarinen 2020; Slaby 2014; Stephan, Walter, and Wilutzky 2014; von Maur 2021). For these situated approaches, moods, emotions, and other forms of affective experience are driven, manipulated, and sus-tained—over multiple timescales—by features of an individual's social and material environment. Accordingly, a full account of emotions, for exam-ple, cannot be given by focusing on the individual alone (i.e., their brain, central nervous system, or even their body and its expressive capacities). This is because the people and things an individual interacts with transform her emotional capacities in fundamental ways. Bodies fit into, and become dynamically "coupled" with, resources in their environments (e.g., tools, technologies, props, practices, other people) in ways that animate and regu-late their emotional experience and behaviour (Colombetti 2016). The ongo-ing input and support of these environmental resources—such as portable music listening technologies, which are powerful tools for on-demand emo-tion-regulation (Krueger 2014a)—allows her to realise otherwise inaccessi-ble forms of experience and expression.

Within this situated context, Slaby and colleagues have recently developed the concept of "affective arrangements" (Slaby, Mühlhoff, and Wüschner 2017). For my purposes, affective arrangements are kinds of atmospheres. They determine both the overall feel and affective tonality of specific locales, as well as what sort of emotion-driven behavioural and social possibilities are present (or absent) within those locales.[1] Affective arrangements are made up of "ensembles of diverse material forming a local layout that operates as a dynamic formation, comprising persons, things, artifacts, spaces, dis-courses, behaviors, and expressions in a characteristic mode of composition and dynamic relatedness" (4). Affective arrangements come in many forms and degrees of intensity. They are nearly ubiquitous in everyday life. And they encompass a range of diverse phenomena: things like corporate work environments (from factories to white-collar work to stock market trad-ing floors), public transportation, street corners, commercial environments (shopping malls, sports stadiums), organisational settings like classrooms, lecture theatres, and worship spaces, as well as both the ritualistic practices that unfold within these myriad spaces (ceremonial regimes like Christmas, Ramadan, election campaigns, birthday parties, baptisms, funerals, etc.), along with the artefacts that support these practices (5).

The key idea here is that affective arrangements shape the form and dynam-ics of how bodies *feelingly* fit into—and come to feel at home within—the part of the world these arrangements carve out.[2] They do so by actively regulating the emotional experiences and behaviour of the individuals who inhabit them. In this way, affective arrangements modulate our agency; they shape how we feel and find our way, including how we find our way alongside others. They have this powerful regulatory impact on us, Slaby and colleagues argue, because affective arrangements are "always in operation, always 'on.' It is the ongoing, 'live' affective relations within the arrangement that constitutes zones of higher relative intensity compared to what is outside" (6). So, dynamic styles of bodily comportment and performative emotional expression appropriate in, say, a raucous pub or restaurant during a night out with friends will not fit into the more sober and serious affective arrangement of a corporate office, mosque, or academic lecture hall. Bodily styles welcome in the former will be actively discouraged in the latter. Accordingly, insofar as we are sufficiently responsive to the norms running through these different arrangements, transitioning from one to the next will actively modulate our agency and affect.

There is another important point to be made. In dynamically shaping our experience and behaviour this way, affective arrangements also shape the character and intensity of the kinds of *interpersonal connections* we develop within these arrangements. Affective arrangements open up—and in ways explored below, close down—social possibilities, ways to connect and share with others.

For example, in her ethnographic work on work cultures, Melissa Gregg observes that increased emphasis on the affective value of "intimacy" in white-collar work relationships has, among other things, led to a willingness to open up one's home and personal life to work colleagues and professional responsibilities. This opening up occurs when we buy into—or are *pressured* to buy into—the importance and practice of full-time connectivity. Gregg found that:

> [T]he social bonds developed between co-workers in the office are a contributing factor in extending work hours. Loyalty to the team has the effect of making extra work seem courteous and common sense ... [even] when loyalties lie not with the organization or even necessarily the job, but with the close colleagues who are the main point of daily interaction.
>
> (Gregg 2011, 85)

Via always-on communication technologies like smartphones, email, and chat apps, we feel an immediate—and *intimate*—sense of connectedness with work colleagues. If we have a question or request, we know that we can almost always get a prompt response. But this felt intimacy does not simply arise from the practical ease with which we exchange information. It also arises, Gregg argues, from the affectively saturated experience of "presence bleed": the experience "whereby the location and time of work become secondary considerations faced with a 'to do' list that seems forever out of control" (2). We respond to email over the weekend, see other colleagues engaging in work activities in real-time (via active email threads, shared documents, etc.) and feel compelled to be a loyal and supportive team member. So, we join in. The experience of presence bleed in this way generates a local instantiation of a workplace arrangement now embedded in our home, no longer confined to the walls of the office.

This local arrangement modulates our agency. Even during our "down time," we feel anxious and compelled to quickly respond to email exchanges before we get left out of decision-making processes. And it also modulates our emotions. We feel urgency to participate, anxiety when we don't (or if we're delayed), and guilt, remorse, and a sense of disloyalty for letting team members down if we take the weekend off while pings on our phone remind us that work is pushing forward without us.

As this example indicates, affective arrangements bring bodies into social alignment with one another (even when those bodies are physically dispersed). They coordinate patterns of shared experience and behaviour. This alignment happens not only because arrangements furnish material resources (e.g., always-on communication technologies) to make this happen. They also generate and maintain normative expectations that *orient* bodies within a given material arrangement (i.e., they determine what to do, and how and when to do it), as well as *signal approval or disapproval* for the way a given body is falling into line—or, conversely, failing to do so (e.g., via giving or withholding promotion or access to VIPs and higher-ups within the organisation).

In this way, arrangements feelingly orient the bodies they come into contact with. This orientation helps bodies fit into their world in different ways. This alignment can be a good thing, such as the way affective arrangements of a nightclub, worship space, or rehab facility can bring bodies into intense modes of connectedness and shared experience that promote well-being. However, as the previous example indicates—and as I examine in more detail below—affective arrangements can also have a *disorienting* effect. They may scaffold the development of unhealthy habits, practices, and forms of self-experience (Maiese and Hanna 2019). One reason for this is that affective arrangements are fundamentally porous. They often bleed into other spaces and arrangements in ways that leave bodies disturbed, restless, unsettled, or on-edge—that is, feeling *not fully at-home* wherever they happen to be (Slaby 2016).

To sum up, the notion of "affective arrangements" is useful here for several reasons. First, it helpfully captures the *dynamic* and *relational* conception of atmospheres I endorsed above. Within affective arrangements like workplaces, museums, gambling casinos, and rehab facilities, individuals are, quite literally, *actively arranged*. Bodies act on their local arrangements and the resources they furnish in order to regulate their experience and behaviour. However, these same arrangements, in turn, act on *them*. As we've seen, the sociomaterial and normative configuration of an arrangement "brings multiple actors into a dynamic, orchestrated conjunction, so that these actors' mutual affecting and being affected is the central dimension of the arrangement from the start" (Slaby, Mühlhoff, and Wüschner 2017, 5).

Second, the notion captures the way that affective arrangements are both *fixed and open-ended*. As the presence bleed example indicates, affective arrangements may begin in one location, such as a specific workplace. However, material resources (e.g., Internet-enabled communication technologies) allow them to expand their ambient range. Affective arrangements can infiltrate a variety of other arrangements; in so doing, they adapt over time and take on new forms. In this way, arrangements are performatively and temporally open-ended. Working from home will involve performative dynamics different from those found between office walls (e.g., responding to email while making dinner or watching a movie on the couch). The normative form and force of a given arrangement can, in this way, remain relatively fixed even while its performative dynamics fluidly evolve and adapt in context-specific ways.

Third, as the previous point indicated, affective arrangements are *nested phenomena*. One arrangement can fit into the general contours of another. A local instantiation of a workplace arrangement, for instance, can spring up at home within the contours of a domestic arrangement. Again, this is because they are fluid and adaptable. Much of their regulative power and potency, therefore, comes from their capacity to infiltrate and spring up within pre-existing arrangements. As we've seen, this can be a good thing. It can help bodies feel at home in the world. A gay Christian, for example, may find comfort and support in certain online communities and spaces that are not available to her in her offline world. This online support may, in turn, help her feel more at home as she navigates intolerance in her offline world. She can access it on an as-needed basis. Online worlds are kinds of affective arrangements that often bleed into and infiltrate our offline experiences and behaviour.[3] But as we've already seen, affective arrangements can also have a negative impact. They can disturb and disorient bodies and dramatically limit or close down possibilities for action and social connection. Affective arrangements have political consequences. I turn to a more focused consideration of these themes now.

## Atmospheres of inclusion and exclusion

I now turn to a more focused consideration of the idea that affective arrangements allow bodies to extend and fit into them in specific ways. I will do so by exploring how certain arrangements are configured in ways that do not afford an easy "fitting into" for certain kinds of bodies. Such arrangements *disorient* certain kinds of bodies and, in so doing, disturb their sense of embodiment, agency, and affect at a deep level. They place bodies in a state of perpetual discomfort. As we'll see, sometimes this discomfort is intentional; sometimes it's not. This focus on disorientation and discomfort will open up yet another way of thinking about how atmospheres do things in and to the bodies that inhabit them. Two examples will be illustrative: Sara Ahmed's critical phenomenology of disorientation and "stopped" bodies and a phenomenological approach to social difficulties in autism.

We spend our days finding our way through different kinds of affective arrangements. As we've seen, many arrangements are designed to make us feel at home in the world. They orient us and help us find our way by guiding and shaping our emotions and sense of bodily agency. Moreover, these arrangements have an important social function. They create what we might refer to as *atmospheres of inclusion.*

Atmospheres of inclusion are designed to bring people together, to coordinate their experiences and behaviour, and in so doing enrich and intensify their sense of interconnectedness. They help us find our way *to* and *with* others. The affective arrangement of a worship space like a cathedral, for instance, may furnish resources for solitary experiences like quiet prayer, reflection, and confession. However, other aspects of this arrangement are deliberately configured to bring people together: from ritualistic and liturgical practices (singing, chanting, reciting the Nicene Creed), to myriad visual markers (icons, paintings, baptismal fonts, gravestones) and organisational features of the nave and other spaces. All of these things collectively remind individuals that they share a common space; they are participating in a historical affective arrangement shaped by a rich tradition of practices and experiences undertaken by people with shared beliefs. In other words, this arrangement helps people find their way through a shared religious form of life.

So, atmospheres of inclusion not only make us feel at home in an individual sense. They also make it clear that we are at home in a world *with others.* Part of their orienting function is to provide cues that ours is a landscape of shared arrangements that bodies can *together* fit into and take shape in. This important social-regulative work that affective arrangements do is often so pervasive and subtle that it is transparent to us. It is easy to overlook and take for granted. However, there are occasions when we do become acutely aware of it and the social possibilities it presents. Often, this happens when we lose access to it. We become aware of the social and regulative character of affective arrangements when, as we try to find our way, we become *disoriented.*

I've already used the term "disorientation" at several points in the discussion. To clarify. I'm using it in a similar way to Ahmed (2007). Disorientation involves a kind of discomfort, a feeling of not knowing how to find one's way. But the experience of disorientation I'm referring to involves more than just getting lost because one lacks the relevant information, such as when we try to navigate a new city for the first time, follow an academic talk when we don't know the relevant literature, or even the disorientation and irritation we feel when puzzling our way through a philosophical problem. These are cases of *epistemic* discomfort. They involve the "irritation of doubt," as Peirce says, that "causes a struggle to attain a state of belief" (quoted in Tschaepe 2021, 2).

Cases of epistemic doubt are an important part of everyday life. They motivate us to do things in order to learn about ourselves and the world more generally. Accordingly, they often have a bodily or behavioural dimension. If I'm unfamiliar with the layout of a city, I can use a map, smartphone, or ask a local for help finding my way. If I don't know how to fix a sink, I can watch YouTube videos or call a plumber. However, in contrast to epistemic discomfort, what I have in mind here—again, following Ahmed—is something slightly different: a richer *felt* sense that one is no longer able to find one's way. And to continue with a theme discussed previously, a central part of its character involves a kind of *bodily discomfort*. More precisely, it is a kind of bodily discomfort that can be present even in the absence of epistemic discomfort.

By "bodily discomfort" I do not mean to suggest that his discomfort is necessarily tied to a specific sensation or part of one's body (e.g., a sore throat, muscle cramp, or irritation that one's shirt doesn't fit quite right). Nor is it necessarily tied to illness—although the discomfort of illness can, like the discomfort I am concerned with, make us feel deeply at odds with both our body and the surrounding environment (Carel 2016; Svenaeus 2019). The bodily discomfort and disorientation I have in mind is rather a global sense of not feeling at home in a particular space—feeling, that is, somehow *bodily out of sync with* or affectively unsettled within the arrangement one happens to occupy at that moment. This experience is multidimensional, and comes in different intensities, degrees, and durations (Tschaepe 2021). For example, a new hire may initially feel bodily out-of-sync with the rhythms, practices, and expectations of the office until they've settled in and "learned the ropes." We may walk into a party alone, quickly scan the room, and suddenly feel bodily disoriented by the sweep of unfamiliar faces.

Despite the differences in how this experience may manifest, this felt loss of at-home-ness is a phenomenological indication that one is no longer finding one's way. If I suddenly realise that I don't know anyone at the party, I may feel my social possibilities abruptly dissolve and be unsure what to do next. Indeed, these experiences may be uncomfortable. But feeling disoriented on the first day of work or when walking into a party full of strangers is a privileged form of disorientation that can be overcome with relative ease (e.g., spending a few days settling into the office; having a stranger come over and introduce you to other partygoers). This is because these arrangements are, for the most part, organised to generate atmospheres of *inclusion*. It may take some time for certain bodies to sort out how best to fit into them. But these arrangements are nevertheless designed to support and enable this process.

In other contexts, however, the discomfort of bodily disorientations can be much more intense and have significant practical and political consequences. Critical phenomenologists are helpful for understanding how so.

Critical phenomenologists incorporate insights from feminist theorists, critical race theorists, queer theorists, decolonial and indigenous scholars, and others to highlight the ethical and political significance of traditional phenomenological debates (Salamon 2018; Weiss, Salamon, and Murphy 2019). Some critical phenomenologists have explicitly drawn our attention to powerful connections between bodily discomfort, disorientation, and the politics of space—that is, the profound, and potentially devastating, consequences of ensuring that certain kinds of bodies (e.g., non-white bodies, queer bodies) are not allowed to comfortably find their way into and through certain kinds of arrangements (e.g., Ahmed 2006, 2007; Fanon 1986; Yancy 2016). This is because certain arrangements are configured to deliberately generate what we might refer to as *atmospheres of exclusion*. Atmospheres of exclusion constrain certain kinds of bodies, hinder their agency and emotions, and in so doing disturb them at a pre-reflective level.

## Non-white bodies and atmospheres of exclusion[4]

Sara Ahmed has developed a rich analysis of the environmental manipulations and bodily dynamics that generate what I'm calling "atmosphere of exclusion." She develops her analysis by drawing our attention to the constitutive relation between body and space. As she repeatedly emphasises, the character of our pre-reflective bodily experience is bound up with space—and more precisely, with specific features of the affective arrangements we inhabit. By "pre-reflective," phenomenologists like Ahmed refer to the fundamental ways we experience our body and its capacities for movement, expression, and action (i.e., our felt sense of agency) (Colombetti 2014). Our body is implicitly present as we perceive and act on the world; it shapes what we experience and how we experience it without our explicit attention from one moment to the next. Ahmed tells us that "the body is habitual insofar as it 'trails behind' in the performing of action, insofar as it does not pose 'a problem' or an obstacle to the action, or is not 'stressed' by 'what' the action encounters ... the habitual body does not get in the way of an action: it is behind the action" (Ahmed 2007, 156). Accordingly, how we feel at home in the world rests on the character of how our bodily experience anchors us in space, including the affective arrangements we inhabit. Ahmed's special contribution is to draw attention to how spaces and arrangements *deprive* us of at-home-ness by placing us in a state of disorientation.

Atmospheres of exclusion generate experiences of disorientation. They hinder certain bodies from finding their way. Ahmed observes that "[f]or bodies that are not extended by the skin of the social, bodily movement is not so easy" (Ahmed 2007, 161). In support of this claim, she develops a phenomenology of "being stopped," as she puts it. Black activism, Ahmed notes, highlights the many ways that policing involves a "differential economy of stopping." Some bodies—mainly non-white bodies—are stopped by the police more than others: e.g., being pulled over while driving, or harassed while trying to enter their own home. But being stopped often occurs in other (i.e., non-policing) contexts, too, such as when non-white bodies are bombarded with racist images or memes in online spaces, followed by suspicious neighbours in gated communities, or passed over for a job despite having equivalent (or better) qualifications than white candidates.

A key insight here is that this stopping doesn't just place *practical* constraints on stopped bodies by depriving them of access to certain things and spaces (although it does). It also has significant *phenomenological* consequences: it induces a perpetual bodily disorientation, a disturbance of that stopped body at a pre-reflective level. This is because the persistent threat of being stopped isn't an abstract or ephemeral thing. It is materially encoded within different affective arrangements designed specifically to unsettle and disorient certain bodies. A stark example is the proliferation of "Whites Only" and "Colored" signs once found above drinking fountains, waiting rooms, toilets, restaurants, and swimming pools across the American landscape well into the 20th century. The prevalence of these signs not only signalled that non-white bodies were not allowed to access the practical resources they offered. They also indicated a lack of *social* possibilities. Non-white bodies were deliberately deprived of possibilities to connect and share, that is, to feel as though they were participatory members of a common history and community.[5]

Of course, there are many more contemporary examples. Some are obvious, such as the 2020 killing of George Floyd, an unarmed African American man killed by the police after allegedly passing a counterfeit bill in a Minneapolis grocery store. Floyd's death under the knee of a police officer inaugurated protests throughout the US and beyond, calling for a change to institutions, practices, and arrangements that systematically devalue and target non-white bodies. However, sometimes this stopping is more subtle; it occurs within arrangements that are supposed to be welcoming and inclusive. These cases reinforce the idea that affective arrangements are porous and nested. Atmospheres of exclusion can spring up within contexts that, on the surface, appear to actively cultivate atmospheres of inclusion.

Ahmed gives us an example. She talks about the experience of walking into a room and experiencing it as a malleable *affective container* (Ahmed 2014, 224). A non-white body can enter into a room full of academic feminists, for instance, and experience a global change in the affective tonality of the room. Even in an academic space one might think would be particularly open and inviting, a non-white body can still have the experience of being stopped. Ahmed quotes bell hooks' description of this experience: "A group of white feminist activists who do not know one another may be present at a meeting to discuss feminist theory. They may feel bonded on the basis of shared womanhood, but the atmosphere will noticeably change when a woman of color enters the room. The white women will become tense, no longer relaxed, no longer celebratory" (quoted in Ahmed 2014, 224).

In this case, the overt reaction of the other bodies (i.e., becoming suddenly tense, less celebratory) makes it clear that the non-white body will not be able to seamlessly fit into the contours of this arrangement. Moreover, part of the bodily disorientation non-white bodies feel is not simply due to the feeling that their presence puts other bodies on edge, that is, that they are somehow affectively out-of-sync with these other bodies and not fully at home. It also arises from the recognition that *they* are the ones who must now work to make others comfortable with their arrival. As Ahmed puts it, "[t]hose who do not sink into spaces, whose bodies are registered as not fitting, often have to work to make others comfortable. Much of what I have earlier called 'diversity work' is thus emotional work" (24). This emotional work extracts a bodily toll.[6]

As these examples indicate, moving through atmospheres of exclusion—pervaded by the persistent threat of being stopped—leaves traces on stopped bodies (Ahmed 2007, 158). These traces are present not only when stopped bodies inhabit acutely threatening spaces, such as being pulled over by the police. It endures when they move on to other spaces, too. As Fanon observes, this is because stopped bodies are perpetually "surrounded by an atmosphere of certain uncertainty" (Fanon 1986, 83). Can I use this toilet? Why did that police car slow down as it drove by? Why did those white feminist scholars stiffen up when I entered the room? Why are the diners at the next table staring at me? Why is this security guard following me as I shop? For both Fanon and Ahmed, no space is entirely free from the threat of being stopped. As a result, "[t]hose who get stopped are moved in a different way" as they find their way through the world (Ahmed 2006, 162). They are disoriented at a pre-reflective bodily level, insofar as they are never allowed to fully extend and take shape within everyday arrangements white bodies take for granted.

Ahmed says that her Muslim name similarly disrupts her bodily experience. It slows her down as she finds her way through the world. This is because her body is continually marked as "could be Muslim," which is, in turn, immediately translated into "could be terrorist." She is, thus, haunted by an atmosphere of exclusion that follows her wherever she goes—simply because she has the "wrong" kind of name. This experience has bodily consequences: "[h]aving been singled out in the line, at the borders, we become defensive; we assume a defensive posture, as we 'wait' for the line of racism, to take our rights of passage away" (163). Ahmed's non-white body is brought into line with other non-white bodies also marked with "terrorist" names. In being singled out and made to wait, government authorities make clear that to be a non-white body in the west "is to be not extended by the spaces you inhabit" (163). Rather, it is to be made to feel continually out-of-sync with—disoriented by and within—those spaces and the atmosphere of "certain uncertainty," the atmospheres of exclusion that pervades them.

Ahmed's analysis is useful for many reasons. Among other things, it provides us with a rich phenomenological account of how certain bodies are made to feel perpetually disoriented by the structure and character of different affective arrangements they inhabit in everyday life. She draws particular attention to the way that affective arrangements, and the atmospheres of exclusion they generate, have political consequences. Equipped with this critical phenomenological framework, we can now turn to a consideration of disorientation and atmospheres of exclusion in autism.

## Autistic bodies and atmospheres of exclusion

How does the previous analysis relate to autism? In short, *autistic bodies are often stopped bodies*. They are not allowed to fully extend into and take shape within the spaces they inhabit—affective arrangements organised primarily around the form of *neurotypical* bodies. This experience of not fitting in, of being hindered from finding their way, can lead autistic persons to experience a kind of pre-reflective bodily disorientation within these arrangements which, in turn, informs and intensifies some of their social difficulties. This claim has significance for understanding the nature of some social difficulties in autism as well as potential intervention strategies. These cases are also helpful for taxonomic reasons. They provide an example of arrangements that can generate atmospheres of exclusion, but which may do so (unlike some of Ahmed's examples) in ways that are unintended by those responsible for them.[7]

First, some brief background. Autistic people often struggle to communicate with others, become attuned to their emotions and intentions, and flexibly adapt to changing social environments. The still-dominant way of thinking about these social difficulties is the neuro-cognitive perspective (Chapman 2019, 422). According to this perspective, autistic differences can be explained by neurocognitive differences found in all autistic individuals. These differences rest on a diminished capacity for *mentalising*, or cognising the existence of other minds, when compared to neurotypicals (Baron-Cohen 1995). This mentalising deficit causes difficulties interpreting and predicting others' behaviour, and smoothly integrating with the shared practices that make up everyday life.

Recently, challenges to this neuro-cognitive perspective have surfaced from a number of fronts. They argue for a more holistic and multidimensional approach. Despite their other differences, these challenges collectively argue that adopting a neurocognitive perspective overlooks key *embodied, interactive, relational*, and *developmental* processes that are partly constitutive of autistic styles of thinking, expressing, and sharing emotions and experiences (Bizzari 2018; De Jaegher 2013; Krueger and Maiese 2018; Roberts, Krueger, and Glackin 2019; Schilbach 2016). Looking at experiences of disorientation and being stopped in autism can help make the importance of some of these processes clearer, as well as what role these relational factors play in shaping some social difficulties.

There is now growing sensitivity to how autistic persons use their bodies to move through the world, express emotions, and respond to the people, things, and spaces around them (Doan and Fenton 2013). Instead of focusing exclusively on cognitive traits, they refocus on distinctive ways autistic persons pre-reflectively experience and live through their bodies as they use their bodily agency to organise sensory information and negotiate shared spaces (Boldsen 2018, Donnellan, Hill, and Leary 2012). Neuro-cognitive perspectives say little about bodily experience in ASD. But understanding the role of the body is crucial for understanding how autistic people find their way through everyday arrangements.

From a neurotypical perspective, ASD styles of embodiment can seem unusual or strange. The timing and flow of their movements may seem somehow off or contextually inappropriate. People with ASD may have an unusual gait or posture. And they sometimes have movements, tics, and habits (e.g., rocking, hand-flapping, spinning, exaggerated gestures, etc.) that neurotypicals find strange. They may also repeatedly shrug, squint, pout or rock back and forth; repeatedly touch a particular object; turn away when someone tries to engage with them; maintain an unusual or inert posture; appear "stuck" in indecisive movements for an uncomfortably long period of time; have trouble imitating actions; or require explicit prompts to perform an action.

These distinct styles of embodiment aren't simply apparent from a third-person vantage point, however. First-person reports suggest that people with autism pre-reflectively experience their body *from the inside* in ways that are different from neurotypical experience. The character of these anomalous bodily experiences shapes their distinctive behaviour which can, in turn, lead to difficulties fitting into neurotypical arrangements.

For example, reports indicate that people with autism often experience difficulties with movements. This includes controlling, executing, and combining movements—from fine motor control, grip planning, and anticipatory movements, to more complex actions like gesturing, reaching for a book, dancing, or negotiating a crowded hallway (Eigsti 2013). Sometimes this feeling results not just from measurable coordination difficulties but also from a *felt* sense of diminished agency and bodily control—including a sense that one's body has a mind of its own, particularly when stressed or overstimulated: "I had an automatic urge to touch my body—rub my thighs or my stomach and chest" (Robledo, Donnellan, and Strandt-Conroy 2012, 6). At other times, individuals with ASD report difficulty feeling their limbs in relation to one another and space (Blanche et al. 2012). This spatial difficulty can make it difficult to smoothly interact with the environment. To cope, some individuals seek sustained deep pressure or joint compression to regain a felt sense of bodily integrity (Leary and Donnellan 2012, 60). Strategies include lying on the floor under a mattress or sofa cushions, jumping on the floor or bed, wearing multiple layers of clothing, banging fists on hard surfaces, or sitting in a plush recliner, bathtub, or swimming pool in order to have the experience of being touched over their entire body.

So, how does all this relate to atmospheres of exclusion? The key point is this: these anomalous bodily experiences can lead people with ASD to feel as though their unique styles of embodiment do not smoothly integrate with neurotypical arrangements, including patterns of interaction and normative expectations comprising these arrangements. Some of the causal factors responsible for these anomalous bodily experiences likely reside within the neurophysiology of the individual. However, some of these factors also appear to be *social*: individuals have the experience of being "stopped" by structures and norm-governed character of neurotypical arrangements. Accordingly, this sense of being stopped feeds into and intensifies aspects of their pre-reflective bodily disorientation when they inhabit and try to negotiate these spaces.

We can let people with ASD describe their own experiences of being stopped, as well as the feeling of bodily disorientation that ensues.[8] One individual says that, "I was sitting on the floor and when I got up after looking at a couple of books, my friend said I got up like an animal does"—and further, that although she is aware that her bodily style differs from those of neurotypicals, she remains unsure of *how* it differs, exactly (Robledo, Donnellan, and Strandt-Conroy 2012, 6). Another says that she will often "lose the rhythm" required to perform actions involving two or more movements, and that "[e]verything has to be thought out" in advance (6), which she is aware gives her movements an excessively stiff and unnatural quality. This felt disconnection both from her own body, along with a sense that she is rhythmically out-of-sync with the neurotypical people and arrangements she inhabits—and judged negatively because of this—cause frustration. It also deepens her sense of bodily disorientation: "I have been endlessly criticized about how different I looked, criticized about all kinds of tiny differences in my behavior ... No one ever tried to really understand what it was like to be me ..." (6). For many people with ASD, negotiating neurotypical arrangements involves negotiating an "atmosphere of certain uncertainty," as Fanon puts it. These arrangements are not set up to accommodate or be responsive to non-neurotypical styles of embodiment and expression. This can lead to the feeling that one is always about to be negatively impacted or judged for not settling into the bodily dynamics of these spaces in a comfortably familiar (i.e., neurotypical) way.

There are many more reports like these (see Leary and Donnellan 2012). They suggest that autistic bodies struggle to extend themselves into arrangements organised around the form, and *norms*, of neurotypicals people and practices. Forms of engagement, expression, and sharing acceptable within ASD forms of life are often actively discouraged and negatively evaluated within neurotypical arrangements. This pervasive resistance gives rise to experiences of atmospheres of exclusion. These atmospheres limit bodily possibilities for people with ASD. Additionally, they shape their feeling of being *bodily stopped*. This resistance might be acutely felt when negotiating the material structure of different neurotypical arrangements such as a noisy, brightly lit lecture hall, restaurant, or retail space that negatively impacts an individual's auditory and visual hypersensitivity. But it can also be felt in different ways within interpersonal contexts, too.

Consider delayed responses in conversation. Autistic people are often thought to struggle with the back-and-forth flow of conversations. Yet, Donnellan and colleagues found that twelve young adolescents with minimal verbal skills, all of whom were labelled developmentally disabled or autistic, could offer competent conversational responses—but only, on average, after 14 seconds of silence (Leary and Donnellan 2012, 57). Most neurotypicals would find this slower-paced rhythm awkward. It would alter the character of that interaction in an unfamiliar way (i.e., for neurotypicals), and they would probably change the subject or leave the conversation altogether.

Consider another conversational example: when someone is asked a question like "Do I look good in this shirt?"[9] An autistic person might see this question as fact-seeking and give an honest and direct answer ("No, you do not"). However, sensitive attunement to the broader context in which it is asked might show that the asker is actually seeking not information but affirmation ("Sure, you look great!"), or at least honest but gentle critical feedback ("Hmm, not bad, but perhaps we can find a more flattering color"). So, a direct and honest answer from an autistic person might be met with confusion, a hurt reaction, and lead to conflict—all of which they may find puzzling and disorienting. Repeated experiences of this sort may discourage them from future engagements. They further intensify the sense that neurotypical arrangements create atmospheres that perpetually exclude them.

Note, however, that this lack of social sensitivity and feeling of fitting in cuts both ways. As McGeer notes, people with ASD may be "blind to our minds, but so too are we blind to theirs" (McGeer 2009, 524). Seeing how so helps to further highlight the spatial origin of some social impairments in ASD. For example, within autistic spaces, it is normal and acceptable for autistics to avoid eye contact when speaking to someone. Within neurotypical spaces, however, people who do this are often seen as deceptive or dishonest. Similarly, neurotypicals may find rhythmic patterns of "self-stimulation" (or "self-stims")—hand-flapping, finger-snapping, tapping objects, repetitive vocalisations, or rocking back and forth, etc.—socially off-putting, and view them as meaningless behaviour. Indeed, treatment programs (often developed with little input from autistic people) have traditionally tried to suppress or eliminate them (Azrin, Kaplan, and Foxx 1973). Yet, for many autistic people, self-stims are embodied strategies for managing sensory information and finding their way. They may use them to refocus and self-regulate when information threatens to be overwhelming (hypersensitivity), or when they require heightened arousal in order to access further information (hyposensitivity). While people with ASD may be actively discouraged from bodily extending themselves via these strategies within neurotypical spaces, they nevertheless have the freedom to do so within autistic arrangements where their meaning and salience is recognised.[10]

The takeaway lesson is that many of the social difficulties autistic people exhibit are *context sensitive*. They are the result of atmospheres of exclusion arising from neurotypical arrangements not adequately configured to accommodate diverse styles of bodily being-in-the-world. These atmospheres are the source of much of the bodily disorientation people with ASD feel in their everyday life. Tellingly, these same social difficulties do not arise when people with ASD inhabit autistic arrangements—atmospheres of inclusion—where these bodily practices are viewed as acceptable strategies for finding one's way. As one autistic person tells us: "If I socialize with other Aspergians of pretty much my own functionality, then all of the so-called social impairments simply don't exist ... we share the same operating systems, so there are no impairments" (Cornish 2008, 158). Reports like these are supported by studies indicating that while high-functioning autistic people may feel anxiety and encounter difficulties interacting with non-autistic people, they nevertheless find their interactions with other autistic persons efficient and pleasurable (Schilbach 2016; see also Komeda et al. 2015). Again, the latter are governed by ASD-friendly norms, expectations, and social possibilities that allow them to bodily extend into those arrangements in ways they cannot when they inhabit many neurotypical spaces.

## Conclusion

I've argued that atmospheres do things. They have a profound regulative power to actively shape experience, behaviour, and forms of social connection. This regulative power, I've argued further, comes from the fact that atmospheres do more than just provide affective colour or texture to the world. They also furnish possibilities: ways to act on the world, ways to fit into the spaces and arrangements we negotiate, alone and with others, throughout everyday life. What I've termed "atmospheres of inclusion" and "atmospheres of exclusion," applied to Sara Ahmed's critical phenomenology of stopped bodies and social difficulties in autism, can show how atmospheres both help and hinder as we find our way.

To be clear, none of the above should be read as suggesting that the ontological focus characterising many ongoing philosophical discussions of atmospheres is a waste of time. It's not. For, despite their ubiquity in everyday life, atmospheres remain an elusive phenomenon. Clarifying their nature, therefore, remains a philosophically useful project. Instead, this analysis should be read as a reminder that atmospheres make a concrete difference in our lives. This is true not just in terms of enhancing our aesthetic and emotional experience. They also allow us to fit into our world, to feel at home in it (or, as we've seen, become disoriented). In other words, they have a profound political and ethical significance worthy of ongoing philosophical attention. Resources from critical phenomenology can help us find our way through some of these issues.

## Notes

1  Slaby and colleagues are hesitant to use the term "atmosphere" for some of the reasons I describe above. Moreover, they say that on some, but not all occasions, the overall affective dynamics that make up affective arrangements can be aptly described as affective atmospheres (ibid., footnote 24). However, it's not clear to me that an affective arrangement, under their characterisation, can *fail* to generate some sort of atmospheric properties. Even an affective arrangement that, for whatever reason, fits together in a discordant or incongruous way will nevertheless still have an overall unifying feel or affective tonality—an atmosphere. So, I will use "affective arrangements" and "atmospheres" in roughly the same way, even if that departs from Slaby et al.'s precise usage.

2  Michelle Maiese (2018) has recently explored similar themes with her rich analysis of the interrelation between social institutions and embodied "habits of mind."

3  Although I don't discuss the Internet here, online spaces can function as affective arrangements. See, e.g., Osler (2020), Krueger and Osler (2019), and Osler and Krueger (forthcoming) for more discussion, including a discussion of why the "online/offline" distinction is increasingly less tenable.

4  The discussion in this section and the next has been adapted and expanded from analysis in Krueger (forthcoming).

5  It is often assumed that these signs were confined to the South. But this is not the case—and some could still be found throughout various parts of the US into the 1970s (Abel 2010).

6  As we'll see, people with autism sometimes describe a similar experience.

7  I here follow the terminological preferences of neurodiversity proponents who, by endorsing identity-first language ("autistic persons") instead of person-first language ("individuals with autism"), deliberately stress the connection between cognitive styles and selfhood (Pellicano and Stears 2011).

8  Chapman (2019) observes that first-person reports of autistic people are often left out of philosophical and psychological discussions of autism (p. 426).

9  This example is taken from Chapman (2019, p. 430).

10  Observations such as these helps explain why the Internet is so important for providing spaces for autistic people to develop online arrangements governed by autistic norms, vocabularies, and styles of expression (Hacking 2009). See Osler (forthcoming) for a phenomenological discussion of how the lived body can enter online spaces and be empathically available to others within those spaces.

## References

Abel, Elizabeth. 2010. *Signs of the Times: The Visual Politics of Jim Crow.* Berkeley: University of California Press.

Ahmed, Sara. 2006. *Queer Phenomenology: Orientations, Objects, Others.* Durham: Duke University Press.

————. 2007. "A Phenomenology of Whiteness." *Feminist Theory* 8 (2): 149–168.

————. 2014. *The Cultural Politics of Emotion*. Edinburgh: Edinburgh University Press.

Anderson, Ben. 2009. "Affective Atmospheres." *Emotion, Space and Society* 2 (2): 77–81.

Azrin, N. H., S. J. Kaplan, and R. M. Foxx. 1973. "Autism Reversal: Eliminating Stereotyped Self-Stimulation of Retarded Individuals." *American Journal of Mental Deficiency* 78 (3): 241–248.

Baron-Cohen, Simon. 1995. *Mindblindness: An Essay on Autism and Theory of Mind*. Cambridge: MIT Press.

Bizzari, Valeria. 2018. "Like in a Shell: Interaffectivity and Social Cognition in Asperger's Syndrome." *Thaumàzein* 6: 158–179.

Blanche, Erna Imperatore, Gustavo Reinoso, Megan C. Chang, and Stefanie Bodison. 2012. "Proprioceptive Processing Difficulties among Children with Autism Spectrum Disorders and Developmental Disabilities." *The American Journal of Occupational Therapy: Official Publication of the American Occupational Therapy Association* 66 (5): 621–624.

Böhme, Gernot. 1993. "Atmosphere as the Fundamental Concept of a New Aesthetics." *Thesis Eleven* 36 (1): 113–126.

————. 2006. "Atmosphere as the Subject Matter of Architecture." In *Herzog and Meuron: Natural History*, edited by P. Ursprung, 398–407. London: Lars Muller Publishers.

Boldsen, Sofie. 2018. "Toward a Phenomenological Account of Embodied Subjectivity in Autism." *Culture, Medicine and Psychiatry* 42(4): 893–913.

Carel, Havi. 2016. *Phenomenology of Illness*. Oxford: Oxford University Press.

Chapman, Robert. 2019. "Autism as a Form of Life: Wittgenstein and the Psychological Coherence of Autism." *Metaphilosophy* 50 (4): 421–440.

Colombetti, Giovanna. 2014. *The Feeling Body: Affective Science Meets the Enactive Mind*. Cambridge, MA: MIT Press.

————. 2016. "Affective Incorporation." In *Phenomenology for the Twenty-First Century*, edited by J. Aaron Simmons and J. Edward Hackett, 231–248. London: Palgrave Macmillan UK.

Colombetti, Giovanna, and Joel Krueger. 2015. "Scaffoldings of the Affective Mind." *Philosophical Psychology* 28 (8): 1157–1576.

Colombetti, Giovanna, and Tom Roberts. 2015. "Extending the Extended Mind: The Case for Extended Affectivity." *Philosophical Studies* 172 (5): 1243–1263.

Cornish. 2008. "A Stranger in a Strange Land: A Journey Through the Social Weirdness of the Neurotypical." In *Asperger's Syndrome and Social Relationships*, edited by Genevieve Edmonds and Luke Beardon, London: Jessica Kingsley Publishers, 151–160.

De Jaegher, Hanne. 2013. "Embodiment and Sense-Making in Autism." *Frontiers in Integrative Neuroscience* 7 (15): 1–19.

Doan, Michael, and Andrew Fenton. 2013. "Embodying Autistic Cognition: Towards Reconceiving Certain 'Autism-Related' Behavioural Atypicalities as Functional." In *The Philosophy of Autism*, edited by Jamie L. Anderson and Simon Cushing, 47–71. New York: Rowman and Littlefield Publishers Inc.

Donnellan, Anne M., David A. Hill, and Martha R. Leary. 2012. "Rethinking Autism: Implications of Sensory and Movement Differences for Understanding and Support." *Frontiers in Integrative Neuroscience* 6 (124): 1–11.

Eigsti, Inge-Marie. 2013. "A Review of Embodiment in Autism Spectrum Disorders." *Frontiers in Psychology* 4 (224): 1–10.

Fanon, Frantz. 1986. *Black Skin, White Masks*. Translated by Charles Lam Markmann. London: Pluto Press.

Gregg, Melissa. 2011. *Work's Intimacy*. Cambridge: Polity.

Griffero, Tonino. 2014. *Atmospheres: Aesthetics of Emotional Spaces*. Burlington, VT: Ashgate Publishing, Ltd.

Griffiths, Paul, and Andrea Scarantino. 2009. "Emotions in the Wild: The Situated Perspective on Emotion." In *The Cambridge Handbook of Situated Cognition*, edited by M. Aydede and P. Robbins, 437–53. Cambridge: Cambridge University Press.

Hacking, Ian. 2009. "Autistic Autobiography." *Philosophical Transactions of the Royal Society of London. Series B, Biological Sciences* 364 (1522): 1467–1473.

Komeda, Hidetsugu, Hirotaka Kosaka, Daisuke N. Saito, Yoko Mano, Minyoung Jung, Takeshi Fujii, Hisakazu T. Yanaka, et al. 2015. "Autistic Empathy toward Autistic Others." *Social Cognitive and Affective Neuroscience* 10 (2): 145–152.

Krueger, Joel, and M. Maiese. 2018. "Mental Institutions, Habits of Mind, and an Extended Approach to Autism." *Thaumàzein* 6: 10–41.

Krueger, Joel. 2014a. "Affordances and the Musically Extended Mind." *Frontiers in Psychology* 4 (1003): 1–13.

———. 2014b. "Varieties of Extended Emotions." *Phenomenology and the Cognitive Sciences* 13 (4): 533–555.

Krueger, Joel, and Lucy Osler. 2019. "Engineering Affect: Emotion Regulation, the Internet, and the Techno-Social Niche." *Philosophical Topics* 47 (2): 205–231.

Leary, Martha R., and Anne M. Donnellan. 2012. *Autism: Sensory-Movement Differences and Diversity*. Cambridge: Cambridge Book Review Press.

Maiese, Michele. 2018. "Life Shaping, Habits of Mind, and Social Institutions." *Natureza Humana - Revista Internacional de Filosofia E Psicanálise* 20 (1): 4–28.

Maiese, Michelle, and Robert Hanna. 2019. *The Mind-Body Politic*. London: Palgrave Macmillan.

Mcgeer, Victoria. 2009. "The Thought and Talk of Individuals with Autism: Reflections on Ian Hacking." *Metaphilosophy* 40 (3–4): 517–530.

Osler, Lucy. 2020. "Feeling Togetherness Online: A Phenomenological Sketch of Online Communal Experiences." *Phenomenology and the Cognitive Sciences* 19 (3): 569–588.

———. 2021. "Interpersonal Atmospheres: An Empathetic Account." University of Exeter. https://ore.exeter.ac.uk/repository/handle/10871/124306.

———. 2021. "Taking empathy online," *Inquiry*, doi:10.1080/0020174X.2021.1899045

Osler, Lucy, and Joel Krueger. Forthcoming. "Taking Watsuji Online: Betweenness and Expression in Online Spaces." *Continental Philosophy Review*

Pellicano, Elizabeth, and Marc Stears. 2011. "Bridging Autism, Science and Society: Moving toward an Ethically Informed Approach to Autism Research." *Autism Research: Official Journal of the International Society for Autism Research* 4 (4): 271–282.

Roberts, Tom, Joel Krueger, and Shane Glackin. 2019. "Psychiatry Beyond the Brain: Externalism, Mental Health, and Autistic Spectrum Disorder." *Philosophy, Psychiatry, & Psychology: PPP* 26 (3): E – 51–E – 68.

Robledo, Jodi, Anne M. Donnellan, and Karen Strandt-Conroy. 2012. "An Exploration of Sensory and Movement Differences from the Perspective of Individuals with Autism." *Frontiers in Integrative Neuroscience* 6 (107): 1–13.

Saarinen, Jussi A. 2020. "What Can the Concept of Affective Scaffolding Do for Us?" *Philosophical Psychology* 33 (6): 820–839.

Salamon, Gayle. 2018. "What's Critical about Critical Phenomenology?" *Journal of Critical Phenomenology* 1 (1): 8–17.

Schilbach, Leonhard. 2016. "Towards a Second-Person Neuropsychiatry." *Philosophical Transactions of the Royal Society of London. Series B, Biological Sciences* 371 (1686): 20150081.

Schmitz, Hermann. 2019. *New Phenomenology: A Brief Introduction.* Mimesis.

Slaby, Jan. 2014. "Emotions and the Extended Mind." In *Collective Emotions,* edited by Mikko Salmela and Christian Von Scheve, 32–46. Oxford: Oxford University Press.

———. 2016. "Mind Invasion: Situated Affectivity and the Corporate Life Hack." *Frontiers in Psychology* 7 (266): 1–13.

———. 2019. "Atmospheres—Schmitz, Massumi and beyond." *Music as Atmosphere: Collective Feelings and Affective Sounds,* edited by Friedlind Riedel and Juha Torvinen. London: Routledge.

Slaby, Jan, Rainer Mühlhoff, and Philipp Wüschner. 2017. "Affective Arrangements." *Emotion Review: Journal of the International Society for Research on Emotion* 11 (1):3–12. doi:10.1177/1754073917722214

Stephan, Achim, Sven Walter, and Wendy Wilutzky. 2014. "Emotions beyond Brain and Body." *Philosophical Psychology* 27 (1): 65–81.

Svenaeus, Fredrik. 2019. "A Defense of the Phenomenological Account of Health and Illness." *The Journal of Medicine and Philosophy* 44 (4): 459–478.

Tschaepe, Mark. 2021. "Somaesthetics of Discomfort: Enhancing Awareness and Inquiry." *European Journal of Pragmatism and American Philosophy* XIII (1): 1–10.

Von Maur, Imke. 2021. "Taking Situatedness Seriously. Embedding Affective Intentionality in Forms of Living." *Frontiers in Psychology* 12 (5999939): 1–14.

Weiss, Gail, Gayle Salamon, and Ann V. Murphy. 2019. *50 Concepts for a Critical Phenomenology.* Evanston: Northwestern University Press.

Yancy, George. 2016. *Black Bodies, White Gazes: The Continuing Significance of Race in America.* Lanham, MD: Rowman & Littlefield.

# Part III

# Aesthetic and political atmospheres

# 7 Shared or spread? On boredom and other unintended collective emotions in the cinema

*Julian Hanich*

## Audience effects in the cinema

Imagine it's a Saturday evening and you are sitting in your favourite multiplex cinema to watch a comedy. Well, let's call it a *wannabee* comedy, because the film, although trying to create a light and cheerful atmosphere, turns out entirely unfunny. In fact, it is impossible for you to honour the filmmakers' intentions even with the faintest of smiles. Most other viewers are not amused in the slightest either. Even the lone viewers in the third and eleventh row, who had occasionally laughed out loud at what all the others consider lame pranks and poor jokes, have slowly calmed down. An atmosphere of fidgety, heavy silence has filled the cinema hall; you and your neighbour and all the other viewers feel bored *individually* and *in parallel* by what the filmmakers have dared to make you go through.

Now, think of another scenario: it's a grey Sunday afternoon and you have gone down to the local arthouse theatre to see the latest film of a serious art-cinema director. You are trying to concentrate on what's going on up there on the screen. It seems entirely obvious that the film intends to radiate an atmosphere of gloomy, heavy momentousness. Yet you find the whole thing pretentious, even ludicrous. Just when you are about to become seriously impatient, somewhere in the dark someone starts laughing. During the next conceited dialogue passage or drawn-out voice-over rumination, you feel encouraged to giggle a bit contemptuously yourself. Others are gradually joining in as well. Before long, the majority of viewers are laughing or uttering acerbic comments. A light-hearted, if sarcastic atmosphere has engulfed parts of the audience, while you and many of the others *share* emotions like amusement and contempt as you go through them *together*. Other spectators, however, feel put off: they are annoyed by what they consider disturbances and acts of disrespect.

On the face of it, these fictive examples are complementary opposites. In the first case, we encounter dead silence when anticipating explosive laughter; in the second case, we have an unexpected audible response where there shouldn't be one. In the first example, viewers remain quietly bored where rumbustious laughter is intended; in the second example, audiences laugh

DOI: 10.4324/9781003131298-8

*Figure 7.1* Cinema audience 1.

out loud about something meant to be experienced in serious silence. In the first scenario, the emotions are *spread* over individual viewers; in the second, the spectators *share* emotions. But the two situations also have things in common. In both cases, the *artist-intended* atmospheres are countered and trumped by *audience-made* atmospheres (Figure 7.1). And in both scenarios, the viewers go through *collective* emotions, albeit of very different kinds.

Admittedly, the two scenarios are pointed, even forced. But the hyperbole serves a rhetorical function: It helps to highlight aspects we can encounter in different degrees also in more mundane movie-theatre situations. If they ring true at least to some extent, they will allow me to pursue two goals.

First, I want to extend a critique that—despite their indebtedness to it—Gernot Böhme and Tonino Griffero have levelled against Hermann Schmitz's notion of atmospheres: that atmospheres can be *actively* produced and that we can even reconstruct a poetics of atmospheres. Böhme calls this *making* of atmospheres "aesthetic work": "We find this kind of work everywhere. It is divided into many professional branches and as a whole furthers the increasing aestheticization of reality. (…) They include: design, stage sets, advertising, the production of musical atmospheres (acoustic furnishing), cosmetics, interior design—as well, of course, as the whole sphere of art proper" (Böhme 2017, 21). However, and here I see a potential to add to Böhme and Griffero's aesthetics myself, atmospheres are not only intentionally created by artists, architects, or designers who want to evoke an atmospheric art experience, but also—voluntarily and involuntarily—by *audiences* who collectively perceive an opera, a theatre performance, a concert or a film. Now, I don't claim that this is a revolutionary insight for those who consider the concept of atmosphere of value in aesthetics and beyond. In fact, it only takes a small step—but this step we still need to take.

Second, I aim to add to the discussion about collective emotions and emotional sharing by introducing the term *spread collective emotions*. Both shared and spread collective emotions are a subclass of collective emotions more widely conceived. But while shared emotions have garnered attention recently, spread collective emotions have flown below the radar. As we will see, unlike the amusement of the viewers in the second example, the boredom an audience collectively endures while watching an excruciatingly tedious comedy is not something they share—at least not in the sense I will define it.

In this essay, I will draw on and extend insights from my book *The Audience Effect: On the Collective Cinema Experience* (2018).[1] The explicit goal of that study was to show what scholars throughout the history of film theory had turned a blind eye to; the fact that the co-presence of other viewers always affects our film experience, for better or worse. This audience effect bears significantly on the *atmosphere* in the cinema hall and the *emotions* we undergo as an audience. When we watch a film in a cinema or another co-viewing situation, we constitute and create a social experience that does not precede this event—it comes alive only through us and, during the film, continuously changes with and because of us. Limiting research to the *dyadic* encounter between a single viewer and the film artificially delimits and distorts the discussion about the film experience. Instead, it's important to realise that the collective constellation is always a *triadic* one between individual viewer, film, and the rest of the audience. In a slogan: Watching a film with others is crucially different from watching a film alone. And this goes, *mutatis mutandis*, also for other aesthetic experiences of a collective kind: pop music concerts, operas, theatre, or dance performances, etc.

To channel, from the beginning, the readers' expectations in the right direction, it may be important to underline that I am writing as a film theorist with a decidedly phenomenological inclination and a strong interest in the philosophical concepts at stake in this volume, but I am not a philosopher. I have long been influenced by Hermann Schmitz's phenomenology and have profited from the work of many scholars involved in this volume. As such, I occupy the curious position of an outsider who feels very much at home. But precisely *as* an outsider I hope to add a useful perspective on phenomena that many readers may be familiar with, but may not have connected to what's at stake in debates about atmospheres and shared emotions. With the help of the concrete case of the cinema experience, I hope to shed light on aspects valuable for the larger philosophical debate as well.

### Change of atmospheres: an active audience

It's well-known that in Gernot Böhme's New Aesthetics atmospheres play a crucial role—atmospheres in all their rich and variegated colours: serious atmospheres, menacing atmospheres, sublime atmospheres, giddy atmospheres, etc. The specificity of atmospheres is best realised when they stand out and we have not yet become used to them: "they are experienced *through contrast*, that is, when finding ourselves in atmospheres that clash with our own emotional state, or when *entering into* them by moving from one atmosphere to another" (Böhme 2017, 168, emphasis added). Accordingly, Böhme distinguishes between *contrastive* and *ingressive* experiences of atmospheres. Unlike Hermann Schmitz, on whom he otherwise relies to an astonishing degree, Böhme believes that atmospheres can be *produced* by relying on the qualities of things—their ecstasies. By using the Greek word *ecstasies*, Böhme wants to indicate that things—including artworks, images, or entertaining films—radiate into space and thereby contribute to establishing an atmosphere: "*Ecstatics* is the way things make a certain impression on us and thus modifying our mood, the way we feel ourselves" (Böhme 2017, 5; see also Griffero 2014, 96–99).

Even though atmospheres are actively produced, not everything produced actually works. It, therefore, makes sense to speak of *intended atmospheres* in case we recognise someone wanted to create an atmosphere but we nevertheless remain unaffected. This can have *artistic* reasons as when the intended atmosphere of light-hearted cheerfulness of a comedy falls flat and fails to have an effect on us. But it can also have *contextual* reasons, for instance, when we decide to watch a dark horror film in broad daylight on our computer screen and the atmosphere of gloomy darkness dissipates rather like the vampire Nosferatu when hit by the first rays of the morning sun.[2]

Another contextual reason for an intended atmosphere to go awry is co-viewers who experience a film differently and thereby create a contrastive atmosphere. Böhme has gestured in that direction as early as 1998: "It shouldn't be forgotten that in our everyday behavior and our ways of life we always co-produce the atmospheres in which we live." He continued with a comment I take as my point of departure: "The everyday interaction as common participation in atmospheres and its communicative creation—that would be another topic" (Böhme 1998, 12, my translation). It is here that I make my moderate intervention: A film's intended atmosphere can conflict with the atmosphere emerging *in* and *from* the audience and the affective affordances of the film remain unexploited or are appropriated for other means.

Consider how in our first scenario the viewers' expectations are crossed out in more than one way. Not only are the spectators expecting the film to *emit* an atmosphere of cheerful hilarity and humour, but they are also anticipating a receptive cinema audience which *resonates* with this atmosphere and responds chuckling and cackling wildly. Neither of the two expectations materialises. Yet a distinction between these two corresponding but different atmospheres make sense; this becomes more tangible when we take a closer look at those two viewers in the third and eleventh row for whom the pranks and jokes weren't all that lame; these two viewers, in fact, considered the film as quite funny. The two pitiable spectators find themselves in a different situation, because the intended *atmosphere of the film* had precisely the affective effect they expected all along. What extinguished their laughter, and thus crossed out *both* their expectations, was the fidgety silent *atmosphere of the cinema hall* constituted by the other viewers. Since the laughter of the two lone spectators did not resonate and find an echo, it eventually faded and their appreciation of the film vanished with it. In this respect, laughing in a group of dead-serious, bored, or otherwise quiet people is rather like screaming in an anechoic chamber: the laughter is sucked up in a void of silence—and dies down.

These spectators must feel like the serious observer, in an example that Hermann Schmitz likes to give, who enters the giddy atmosphere of a party: She is well aware of the giddiness that surrounds her, but doesn't feel giddy herself but rather sad and pensive (Schmitz 2003, 251). With Böhme we could say that the affordances of the comedy—their ecstatic qualities—were merely kept "in latency" (Böhme 2019, 166) for the two spectators and the affective effect has been hampered by the hostile surrounding atmosphere. Had they watched the comedy alone at home, they might have enjoyed it more. And, at this point shifting our focus from atmospheres to emotions, they would not have gone through the collective emotion of boredom.

## Shared and spread collective emotions

Before I can say more about collective boredom in the cinema, I first need to introduce a conceptual distinction crucial for the following discussion—a distinction between *collective* emotions, *shared* emotions, and *spread* emotions. Following Christian von Scheve and Sven Ismer, I prefer a broad definition of collective emotions and consider them as "the *synchronous convergence in affective responding* across individuals towards a specific event or object" (2013, 406, original emphasis). The definition is wide-ranging because (a) collective emotions and individual emotions do not have to differ qualitatively from one another, (b) face-to-face encounters or other forms of co-presence are not required for the synchronous convergence, and (c) individuals don't have to be mutually aware of each other's emotions. For this broad understanding of collective emotions, it suffices that individuals appraise an event in similar ways, share appraisal structures or concerns, and converge in terms of emotional response. As an example, von Scheve and Ismer refer to a traffic jam: the drivers appraise the situation as obstructing their goals; they have limited potential to cope with the situation; and they share the concern that they might arrive late at their destination. This leads to a synchronous convergence of anger or frustration, but encapsulated in their cars, the drivers know very little about each other's emotions and affective expressions.

In my view, shared emotions—or what sometimes also goes by the name of "emotional sharing" (Zahavi 2015; Thonhauser 2020), "shared feeling" (Schmid 2008), or "feeling-in-common" (Max Scheler 2008 [1923])—are a particular type or subclass of collective emotions. Together with *emotional contagion* and *feeling together*, shared emotions constitute the three most common types of *affective we-experiences* in the cinema (see Chapter 6 in Hanich 2018). Drawing on the work of philosophers like Hans Bernhard Schmid (2008; 2014), Dan Zahavi (2015), and Mikko Salmela (2012; 2014), I argue that individuals—such as film viewers in a cinema—share an emotion when four necessary and jointly sufficient conditions are fulfilled (for a similar, albeit slightly different account, see Thonhauser 2020).

First, spectators share an emotion when they experience the *same kind of emotion*. Their emotions may not merely be similar, and they must by no means be dissimilar. While this might sound obvious, it is important to underscore that not every affective we-experience is based on the same kind of emotion. When viewers experience the affective we-phenomenon of *feeling together* they go through different, albeit *matching* emotions (Hanich 2018, 178–181; on feeling together, see also Sánchez Guerrero 2016).

Second, the spectators' emotion must also be directed towards *the same intentional object*. Again, this might sound trivial, but we can easily imagine cases in which viewers are simultaneously amused by very different things: while some are amused by how ridiculously pretentious the film is (as in my second fictitious scenario above), another viewer is amused because she has just exchanged a joke with her neighbour and yet another one has received a funny GIF on his smartphone. This also implies that when two viewers share an emotion, they both *immediately* respond to the shared intentional object, and not to each other's *response* to that object. This distinguishes shared emotions from emotional contagion, where the emotion causally depends on someone else's emotions. And it also sets shared emotions apart from affective forms of empathy and sympathy where togetherness is mediated as well: When I feel *with* you (in empathy) or *for* you (in sympathy) I do it, in a sense, *because of* you. In both cases, the other individual's emotion is the object; my response is mediated, not immediate.

Third, for the same emotion to be shared, some form of *mutual awareness* is necessary. Thus, a coincidental case of qualitatively identical emotions running in parallel must be ruled out. When we share an emotion, I must have at least some peripheral idea that you experience the same emotion as I do and that you know that I know it. As Thonhauser puts it (2020, 208): "individuals who are involved in emotional sharing are aware of each other as co-subjects of that affective experience." However, this does not imply that we have to actively focus upon our mutual awareness when sharing an emotion; it can remain at the fringe of consciousness. Nor does the requirement of mutual awareness imply a strong truth claim: I can be wrong about your emotion just as you can be wrong about mine.

Fourth and lastly, sharing an emotion comes with a certain *loss of distance and individuality* and, thus, an experience of some form of *phenomenological closeness.* Dan Zahavi (2015, 90) also speaks of an "affective bond" or "unification." Only in this case would it be legitimate to say that the sarcastic amusement you and I and all the others go through when we laugh about a preposterous film is *our* amusement. Shared emotions derive from emotions that "open" us to others and even "connect" us to them. Here we can expect different *degrees* of felt closeness and distance and hence, different degrees of sharedness of an emotion (see also Salmela 2012). These degrees of closeness and distance may have to do with the *intensity* of the emotion, but they can also depend on the *kind* of emotion.

This leads us directly to the problem of what a *spread* collective emotion is. Dan Zahavi, in an important article entitled "You, Me and We: The Sharing of Emotional Experiences," has asked, merely in passing, if all emotions can be shared in the same way (2015, 98). In a direct response, Zeyne Okur Güney (2015, 105) has expressed her doubts: "when experiencing emotions such as hate, envy, jealousy, shame, or anger, the distinction between self and other is strongly manifest, whereas in compassion, love, or sympathy it diminishes" I agree: not all emotions allow for a loss of distance and an affective phenomenological closeness to others. Thus, even though both are a subclass of collective emotions, a crucial difference between shared collective emotions and spread collective emotions remains the necessary phenomenological closeness. While it is felt (however, mildly) between those who share a collective emotion, it is missing in spread emotions. In other words, spread collective emotions do not count as an affective *we*-experience. Experiencing a spread collective emotion rather implies that we all have the same immediate emotional response to a shared object or event and are mutually aware of it to some degree, but nevertheless feel individuated and hence (somewhat) detached from each other.

An interesting case in point is collective embarrassment. Imagine a film screening in which, all of a sudden, a very explicit sex scene or even a hardcore pornographic shot appears on the screen. It's not a secret that showing unsimulated sex has become *de rigueur* among art cinema directors like Catherine Breillat, Larry Clarke, Michael Winterbottom, Lars von Trier, Abdellatif Kechiche, Ulrich Seidl, Radu Jude, Gaspar Noë, Bertrand Bonello, and many others.[3] It is not unlikely that this scene can lead to a flash of embarrassment (depending of course on a number of contextual factors: for instance, is it a university screening with students, colleagues, and some superiors, a private gathering at home with a few close friends or a public cinema screening with anonymous co-viewers?). But for the sake of the argument, let's assume that all viewers undergo a moment of embarrassment: it is difficult to imagine that members of the audience suddenly feel more unified or connected to each other. Embarrassment simply does not seem to be the kind of emotion that lends itself to an affective bond in a cinema setting. Thus, the viewers do not have this embarrassing experience together, but in parallel; to them, it only involves for-*me*-ness, not for-*us*-ness. Hence, it would feel incorrect to call the embarrassment you and I and all the others go through when confronted with a pornographic scene, *our* embarrassment.

To be sure, we don't have to decide *ex cathedra* which emotions allow for phenomenological closeness and hence shared emotions, and which ones don't. There may well be emotions that can allow for both and depend on the social context and the type of co-viewers you watch the film with. An example could be moments of being *sadly moved* by a film (on the emotion of being moved, see Kuehnast et al. 2014; Menninghaus et al. 2015; Cova and Deonna 2014; Deonna 2020). On the one hand, when a film moves an entire cinema to tears and the audience, thus, experiences a collective emotion, the individual viewers need not necessarily share the emotion: Although at this moment the entire audience may be sadly moved, not all viewers have to experience it in a *we*-mode. Some may well go through the emotion in an *I*-mode, as if surrounded by an individualising bubble. On the other hand, some pockets of the audience—say, a group of close friends or a mother and her daughter—may share tears together and feel phenomenologically close to each other (on shared weeping, see also Hanich 2018, 240–242). This shows us that within a given audience, collective emotions can be both shared (by some) and spread out (for others). The latter is arguably the case in collective boredom, such as that experienced by the audience in our first scenario.

## Involuntary boredom as a spread collective emotion

Boredom is a widespread negative emotion that has raised considerable interest among philosophers (e.g., Heidegger 1995 [1929/30]; Neu 2000; Elpidorou 2018). But boredom is also an aesthetic emotion and as such has garnered a fair amount of attention in film studies (e.g., Misek 2010; Richmond 2015; Çağlayan 2018; Quaranta 2020; Ferencz-Flatz forthcoming). The reasons for this spark of curiosity among film scholars are plentiful: the rise of interest in emotions; the growing attention to the phenomenology of film experience; and the ascent of slow cinema, by directors like Béla Tarr, Tsai Ming-liang, Lav Diaz, or Pedro Costa, as a vital aesthetic force in global cinema. However, what these studies have sidestepped is precisely the *collective* boredom we are interested in here.

Let us, therefore, hark back once again to my first fictitious scenario of the spectators who are not only underwhelmed by the wannabee comedy but flat-out bored. Following Heidegger's classification of three types of boredom, we can immediately identify it as an example of his first category: "being bored with something" (*Gelangweiltwerden von etwas*).[4] When an atmosphere of booming, but fidgety silence calmed down even the two viewers who were initially laughing out loud, it drowned their enthusiasm, thus changing their emotions from amusement to boredom. With Heidegger (1995 [1929/30], 103), we can say that the film leaves them *empty* because it offers them nothing.

Boredom's emptiness—its *unfulfilledness*—is the opposite of "a fulfilled time." We, therefore, find ourselves in a situation that foregrounds time and its pace.[5] In boredom, time moves too slowly, lingers, seems to stand still. It becomes obtrusive, and we experience the situation as heavy and stultifying. Heidegger (1995 [1929/30], 97) talks appositely about *das Lastende und Lähmende*, the burdensome and paralysing. As a way out, we seek an occupation to fill the emptiness of time. Since there is little else we can do, we take useless action: We look at our watch; we check our cell phone for new messages; we move in our seats and look around; we flee into daydreaming to play out an alternative movie in our private theatre of the mind. All this to bridge the emptiness between the now and what is our yearned-for goal: the end of the film, or at least the beginning of a more entertaining scene. Inasmuch as all viewers are bored with the supposedly funny comedy and, thus, have converged synchronously in affective responding, they undergo a *collective* form of boredom.

But, and this is crucial for my argument, in boredom the film also seems to *abandon us to ourselves* (Heidegger 1995 [1929/30], 103). Or, as Andreas Elpidorou puts it: "Bored individuals experience *a withdrawal from their environment* and cannot identify with what the environment is offering them" (2018, 460, emphasis added). Thus, the experience of boredom arguably comes with a certain distanciation from the world, an isolating encapsulation, even in the close proximity of others. In fact, we can be bored even when doing something *with* others like playing a boring card game, carrying out a monotonous task or, indeed, watching a tedious movie together. In the following quote Lars Svendsen (2005, 112) uses the term "mood" instead of emotions, but boredom can be both a mood and an emotion for him (on boredom as an emotion, see also Elpidorou 2018): "Experiences become possible by virtue of moods that are suitable for it. Certain moods may incite sociality (e.g., joy), whereas others are more likely to lead to loneliness (e.g., boredom)." This is not to say that boredom always comes with a feeling of loneliness, but it points to our tendency to not experience it in a we-mode. Bored viewers in the cinema lack the necessary closeness to each other that comes with emotional sharing. They are not bored *together* in an emphatic sense, but bored *for themselves* and *next to each other*.

If they were feeling close, they would not undergo the very lack of meaning and emptiness so characteristic of boredom—they would have already chased boredom away. In this respect it's worthwhile mentioning that proponents of a functional account of boredom like Andreas Elpidorou underline a specific purpose of boredom: it motivates us to pursue a new goal when the current one ceases to be satisfactory, attractive, or meaningful. As such, it propels us to follow strategies to re-establish meaning and pursue *pro-social intentions* (Elpidorou 2020, 1). Similarly, and with reference to Schopenhauer, Svendsen (2005, 172) claims that boredom can lead to sociality precisely as a diversion from boredom. This means that we sometimes aim to overcome boredom's withdrawal and distanciation from the world and others precisely by seeking out others. This seems to me another clue: boredom in the cinema is a spread collective emotion, not a shared one. We can even find a source of relief and even pleasure when discovering, during the film or afterwards, that others judge it as equally tedious and lame. But again, this does not imply we share the emotion of boredom because the moment we have reached out to others and discovered the relieving fact that others were equally bored, boredom has disappeared, if only temporarily.

## Unintended laughter and subgroups in the audience

In the final section, I shall now turn to my so far neglected second scenario and draw attention to laughter as a form of emotional expression that can have a strong audience effect on atmospheres and shared emotions. I will concentrate on instances in which laughter—as a straightforward emotional expression of *being amused*, but also of *contempt*—can create a light-hearted or hostile atmosphere that was neither existent beforehand nor intended by the filmmakers. The scenario will also help to underline how volatile atmospheres and shared emotions can be. Not least, it will show that audiences are often anything but coherent groups, but can consist of protean-like subgroups.

Laughter has a centrifugal spatiality outward and implies a transcendence of the self, in the sense that one *ex-plodes, ex-hales,* and, thus, *ex-presses* a sound from inside out and forward into a space often shared with others (Figure 7.2). As we have seen, in their accounts of atmospheres Gernot Böhme and Tonino Griffero often use the evocative term "ecstasies." Laughter, with all its eruptive, outward-moving, and ex-haling characteristics, has exactly this ec-static quality, too, which can influence and even change an atmosphere. In the following, I will—predominantly but not exclusively—deal with what I call *conversion laughter* (for a typology of ten kinds of laughter in the cinema, see Hanich 2018, 193–207). Here the expression of amused or derisive laughter occasions the *evaluative transformation* of a film that was intended to be serious (or sentimental or scary) and is instead being laughed *at*. In these cases, we can witness a gradual change of atmosphere in the cinema hall from serious to light-hearted, or contemptuous and aggressive. Monica Vitti, no less, once told how the audience turned the premiere of Michelangelo Antonioni's *L'Avventura* (1960) at the Cannes Film Festival into a deeply hurtful experience for her. Afterwards she was "crying like a baby." What had happened? "[The viewers] were laughing at the most ... most tragic sequences, those that we had sweated the most over and believed the most in. And this went on throughout the projection."[6]

*Figure 7.2* Cinema audience 2.

However, for my claims about conversion laughter to sound convincing it is important to remind ourselves that laughter does not always have to be strongly eruptive and quasi-automatic. There are instances in which we respond in far more controlled and active ways. While we are often passively "done" by laughter, we sometimes also actively "do" laughter. Laughter, in other words, can take the abrupt form of a bursting explosion, but also the gradual form of a melting erosion (Prütting 2013, 1554). The Hermann-Schmitz-inspired scholar of laughter Lenz Prütting, therefore, suggests a polar continuum: At one end of the spectrum, we find a bursting, overwhelming, quasi-automatic laughter with a maximum loss of autonomy and a minimum of self-assertion. In-between, there are forms of laughter of low intensity in which the viewer's loss of autonomy and his or her self-assertion are balanced equally. At the other end of the spectrum, we can locate fully controlled forms of laughter which imply only a minimum loss of self-control. As an example, we could cite cases when we cognitively *understand* that something is meant to be funny and appreciate this intention with restrained laughter, even though we don't find the scene funny at all. Here the active control is high and we could have just as well inhibited the laughing response and remained silent. Arguably, another case of controlled laughter is the conversion laughter that modifies a film's intended atmospheres, emotions, and meanings.

In general, laughter can function like a performative *value judgment* and evaluate its object along a vertical up-down axis: (1) *Laughing down at* as a smug sign of superiority; (2) *laughing with* as a sign of equality where one recognises a film or its maker as on an equal level as oneself; (3) *laughing up at* as a subversive act that wants to turn an inferiority position upside down (see also Prütting 2013, 1865/1866). In our case, the first and the last type of evaluative laughter are crucial: laughing down at and up at.

On the one hand, the transformative act of laughter can have a benign-humorous quality. The viewer humorously and ironically transforms a serious film with a light-hearted, campy, tongue-in-cheek change of perspective into something worthy of laughter. Think of the humorous cult surrounding thrash films or failed horror movies such as *Troll 2* (1990) by Claudio Fragasso, Tommy Wiseau's *The Room* (2003), *Birdemic: Shock and Terror* (2008) by James Nguyen or films by Ed Wood like *Plan 9 from Outer Space* (1959). In cases like these film scholars speak of *so-bad-it's-good cinema* and cite the director's incompetence, but also the temporal and cultural distance as reasons for the audience's humorous response (Smith 2019). These films are *intended* to create an atmosphere of serious sentimentality, uncanniness or scariness, and elicit concrete emotions like being-moved, horror, and terror. However, through a humorous change of perspective some viewers judge them as hilarious and make this audible to the rest of the audience. From its position of superiority, the cult-movie audience laughs *down* at the film, but its benign-positive laughter at the same time appreciates the object and *pulls it up*, as it were, from its low cultural status.[7]

On the other hand, evaluative transformations via laughter can also have a hostile and contemptuous tone. In my second fictitious scenario, the arthouse film and its director seem to flaunt an air of superiority, or at least a high-minded authority. In their overt seriousness, they put themselves on a high cultural pedestal, but this superior position does not seem rightfully earned and the film, therefore, comes across as rather ludicrous. Inspired by the so-bad-it's-good category, we might speak of the *so-pretentious-it's-ridiculous film*. Viewers usually discover the unjustified authority on the level of the film, but their rejection can also be fuelled by promotional and publicity materials, including highfalutin interviews of the director (McLean 2013, 152). Here laughing—Henri Bergson famously ascribed a punishing effect to it—assumes a corrective function: The viewers' laughter violently shakes the pedestal and, in an act of devaluation and degradation, dethrones what was put there undeservedly. As a kind of acoustic, non-verbal speech-act, it implies a negative judgment of taste that evaluates the film as overly pretentious. Since one would rarely expect this kind of laughter alone at home, I consider it a hostile signal to *other viewers*—a signal that communicates an evaluation close to a grammatical utterance like, "What pretentious nonsense!" or "How lame!"

Let's assume that two or three viewers have—rather actively and intentionally—initiated this type of revaluation of the intended atmospheres and emotions. Some viewers, who feel similarly, might consider this as an invitation to join in; they actively laugh *up* at the film and try to subversively *pull it down* too. At this point, two other forms of laughter might follow. The first one—I call it *mimicry laughter*—implies a more or less active form of laughing-along-with out of conformity or solidarity. Here viewers mimic the laughter of the others because it either seems too *authoritative* or too *convincing* as a form of evaluative transformation to reject support. The second type of laughter that might follow is *contagion laughter*. In this case, other viewers might be *passively* pulled into laughing together with the initiators in an almost involuntary response to the infectious character of laughter. In this case, laughter is not an act of mimicry; the laughter occurs simply because other people have laughed in an infectious way. Yet also in this case, we might experience a considerable change of atmosphere.

All of this might sound rather schematic and too intentional at times. However, the three steps—from intentional conversion laughter, to more or less active mimicry laughter, to passive contagion laughter—are meant as a mere analytic dissection of what can happen almost simultaneously and in various parts of the audience. Moreover, I certainly don't claim that all of this is collectively orchestrated or deeply thought through; the phenomenon can occur spontaneously and with very limited intentions. Since laughter is such an eruptive, acoustic phenomenon, it is hard to ignore, especially against the background of an otherwise silent auditorium. While the film may attempt to establish an atmosphere of high seriousness (or melodramatic sentimentality, horrifying scariness etc.), the laughter easily disrupts and transforms it. Isn't this exactly a case of "everyday interaction as common participation in atmospheres and its communicative creation" (Böhme 1998, 12)?

This leaves us with the question of shared emotions in our second scenario. Laughter, as an expression of a shared emotion of amusement or contempt, can be an obvious means of making viewers *mutually aware* of each other and phenomenologically "uniting" them (however, briefly).

Let's assume that a group of viewers—or even the entire auditorium— laughs in an amused or derisive way about a film meant to be dead-serious, sentimental, or scary. This laughter is equivalent to a physical outburst that expressively exhales a we-statement. The audience expresses a shared *evaluation*: "This is unintentionally funny or ludicrous *for us!*" But it also voices a shared *emotion*: "We are feeling amusement or contempt *together!*" For a brief moment, those spectators who join in the laughing collective give up their self-control and allow a temporary diffusion of the rigid boundaries of individuality characteristic of what Hermann Schmitz calls *personal emancipation*.

But obviously this collective expression by no means implies that the emotion they share is always shared by the *entire* audience. To drive home this point, I will round off my essay by underlining seven ways in which laughter can affect an audience and create various forms of social relation.

First, our collective laughter can bring about an affectively close *we*, where I laugh with you, you laugh with me and we all laugh with each other.

When laughing together about something ridiculous or contemptible, our subjective social distances decrease, or even disappear for a brief moment in time. We may feel, as it were, centripetally pulled together. This affectively close "we" does not know an "Other" in the auditorium: All viewers feel amused or contemptuous and there is no experience outside to this "we."

Second, laughter can also result in a more oppositional affective audience interrelation of *we-thou* (where the "thou" indicates the second-person singular). Think of a group of friends who were persuaded by one of them to watch our overly pretentious arthouse film, a film she has seen on an earlier occasion and finds particularly thought-provoking and moving. Unlike her, however, the others soon realise they collectively detest the film. They use their—somewhat sadistic, derisive—laughter to spoil the film for her and thus create a confrontational stance that puts them into opposition to and at a felt distance from her. While *they* metamorphise into a momentary "we," they simultaneously try to box *her* into the position of a "thou."

Third, laughter may also evoke an antagonistic experience of *we-ye* (where the "ye" refers to the second-person plural). In this case, laughter pits factions of the audience against each other. For instance, those who derisively laugh together about the—for them: overly pretentious—film might actively position themselves against those silent others who (presumably) find it moving, just as much as these silent others might feel opposed by the aggressively laughing segment. As a consequence, a couple or a group of friends—who are all fans of the arthouse director's work—might exchange glances and signal to one another their anger. In that moment, as a group, they share an emotion that pits them against those they find disrespectful.

Fourth, laughter may evoke a less specific *we-they* experience, in which "they" implies a rather undefined background, a vague outside of the group that remains indeterminate and is not reflected upon. For instance, those who find a pretentious film ridiculous may not intend their laughter to create any opposition but simply express their contempt together. Sharing their humour allows both the creation of a feeling of togetherness and marking the boundaries of the group without making the outside of the group a defining factor (as in the previous two cases).

Fifth, an individual viewer may feel opposed to a particular spectator—an experience we could call the *I-thou* mode (second-person singular). Think of a grumpy husband who laughs derisively down at the film in order to spoil his wife's deeply felt experience; or a viewer moved to tears, getting annoyed by another viewer's degrading laughter. In these moments, the viewers do not experience any shared emotion about the film together, but rather direct their distancing emotions—such as anger or frustration—*at* the other.

Or, and this is my sixth category, a single viewer may feel rejected from or opposed to a bigger group or even the entire audience. We could call this the *I-ye* mode (second-person plural). A case in point would be the mocking laughter of a single viewer who tries to evaluate the preposterous film negatively but no one else follows. At this point, she may unwittingly feel excluded from the rest while embarrassment wells up in her. Inversely, cowering in his seat we find a young boy who is deeply scared by *Birdemic* or *Plan 9 from Outer Space* and feels isolated and excluded because all the others judge the film as hilarious and laugh at it.

Seventh, and finally, we can imagine an *I-they* experience in which the rest of the audience remains a vague backdrop not reflected upon, comparable to the *we-they* mode above. Here we could think of a viewer who snickers sneeringly to himself about how ridiculously pompous the movie comes across, without targeting the rest of the audience in any pronounced way. The viewer neither feels positively individualised nor negatively isolated because the rest of the audience remains a mere background.

This discussion has tried to accentuate that spectators can—willingly or unwillingly—transform atmospheres in the cinema hall, change a film's intended effect, coagulate into groups sharing emotions, dissolve into clusters whose collective emotions are spread out or feel individually distanced from others. Needless to say, there is much more that can be said about audience effects, but one thing has hopefully emerged: in our discussions about atmospheres and emotions in aesthetics, audiences represent a force to be reckoned with.

## Notes

1 I thank Edinburgh University Press for allowing me to reuse material previously discussed in Chapters 6 and 7 of my book.
2 In film studies, the neo-phenomenological term "atmosphere" has not left a big mark yet (in Brunner, Schweinitz, Tröhler 2012, for instance, it doesn't play a role). For film scholars, it is more common to talk about moods and *Stimmungen* (see Smith 2003; Plantinga 2012; Sinnerbrink 2012).
3 See for instance this Wikipedia list: https://en.wikipedia.org/wiki/Unsimulated_sex
4 For the viewers in the third and eleventh row, in case they were able to resist the collective boredom and enjoy the movie, the film experience afterwards might have turned out to be an example of the second type of boredom: "being bored with something" (*das Sichlangweilen bei etwas*). For Heidegger, this is the case when someone has successfully whiled away time, but later on realises that the whole experience was empty after all. For an interpretation of mainstream entertainment film as a "cinema that kills time," see Quaranta 2020, 10–13.
5 As Lars Svendsen explains (2005, 127): "Time is usually transparent—we do not take any notice of it—and it does not appear as a something. But in our confrontation with a nothing in boredom, where time is not filled with anything that can occupy our attention, we experience time as time."
6 The interview with Vitti can be found on the YouTube channel of the Criterion Collection: https://www.youtube.com/watch?v=QEJuAnG0ND0. The English translation from the Italian original was taken from the video itself.
7 In reality, the cult-movie laughter is not always benign. It can also be a cruel sign of derision, as Iain Robert Smith (2019, 713) warns us: "It is dangerous [...] to simply treat this camp engagement with 'so bad it's good' cinema as harmless laughter at failed intention or to treat it 'objectively' as if [...] questions of cultural and ethnic power are not involved."

# References

Böhme, Gernot. 1998. *Anmutungen: Über das Atmosphärische*. Ostfildern: Edition Tertium.

Böhme, Gernot. 2019. *Leib: Die Natur, die wir selbst sind*. Berlin: Suhrkamp.

Böhme, Gernot. 2017. *The Aesthetics of Atmospheres: Ambiances, Atmospheres and Sensory Experiences of Spaces*. New York: Routledge.

Brunner, Philipp, Jörg Schweinitz, and Margrit Tröhler. 2012. *Filmische Atmosphären*. Marburg: Schüren.

Çağlayan, Emre. 2018. *Poetics of Slow Cinema: Nostalgia, Absurdism, Boredom*. Cham, Switzerland: Palgrave Macmillan.

Cova, Florian, and Julien A. Deonna. 2014. "Being Moved." *Philosophical Studies* 169, no. 3: 447–466. https://doi.org/10.1007/s11098-013-0192-9.

Deonna, Julien A. 2020. "On the Good That Moves Us." *The Monist* 103, no. 2: 190–204. https://doi.org/10.1093/monist/onz035.

Ferencz-Flatz, Christian. (Forthcoming.) "Depth of Experience: On the Value of Profound Boredom in the Cinema." In *What Film Is Good For: On the Ethics of Spectatorship*, edited by Julian Hanich and Martin Rossouw. Oakland: University of California Press.

Elpidorou, Andreas. 2020. "Is Boredom One or Many? A Functional Solution to the Problem of Heterogeneity." *Mind and Language* 36: 491–511. https://doi.org/10.1111/mila.12282.

Elpidorou, Andreas. 2018. "The Bored Mind Is a Guiding Mind: Toward a Regulatory Theory of Boredom." *Phenomenology and the Cognitive Sciences* 17, no. 3: 455–484. https://doi.org/10.1007/s11097-017-9515-1.

Griffero, Tonino. 2014. *Atmospheres: Aesthetics of Emotional Spaces*. Translated by Sarah De Sanctis. Farnham Surrey, England: Ashgate.

Güney, Zeyne Okur. 2015. "Collective Affectivity as a Flux of You, Me, and We: Comment on Zahavi's 'You, Me, and We: The Sharing of Emotional Experience'." *Journal of Consciousness Studies* 22, no. 1–2: 102–6.

Hanich, Julian. 2018. *The Audience Effect: On the Collective Cinema Experience*. Edinburgh: Edinburgh University Press.

Heidegger, Martin. 1995 [1929–1930]. *The Fundamental Concepts of Metaphysics: World, Finitude, Solitude*. Translated by William A MacNeill and Nicholas Walker. Bloomington: Indiana University Press.

Kuehnast, Milena, Valentin Wagner, Eugen Wassiliwizky, Thomas Jacobsen, and Winfried Menninghaus. 2014. "Being Moved: Linguistic Representation and Conceptual Structure." *Frontiers in Psychology* 5. https://doi.org/10.3389/fpsyg.2014.01242.

McLean, Adrienne L. 2013. "If Only They Had Meant to Make a Comedy. Laughing at *Black Swan*." In *The Last Laugh: Strange Humors of Cinema*, edited by Murray Pomerance, 143–161. Detroit: Wayne State University Press.

Menninghaus, Winfried, Valentin Wagner, Julian Hanich, Eugen Wassiliwizky, Milena Kuehnast, and Thomas Jacobsen. 2015. "Towards a Psychological Construct of Being Moved." *Plos One* 10, no. 6: e0128451. https://doi.org/10.1371/journal.pone.0128451.

Misek, Richard. 2010. "Dead Time: Cinema, Heidegger, and Boredom." *Continuum* 24, no. 5: 777–785.

Neu, Jerome. 2000. "Boring from within: Endogenous versus Reactive Boredom." In *A Tear Is an Intellectual Thing: The Meanings of Emotion.* New York: Oxford University Press, 104–116.

Plantinga, Carl. 2012. "Art Moods and Human Moods in Narrative Cinema." *New Literary History* 43, no. 3: 455–475.

Prütting, Lenz. 2013. *Homo Ridens: Eine phänomenologische Studie über Wesen, Formen und Funktionen des Lachens. In drei Bänden.* Freiburg, Breisgau: Alber.

Quaranta, Chiara. 2020. "A Cinema of Boredom: Heidegger, Cinematic Time and Spectatorship." *Film-Philosophy* 24, no. 1: 1–21. https://doi.org/10.3366/film.2020.0126.

Richmond, Scott C. 2015. "Vulgar Boredom, or What Andy Warhol Can Teach Us About Candy Crush." *Journal of Visual Culture* 14, no. 1: 21–39. https://doi.org/10.1177/1470412914567143.

Salmela, Mikko. 2012. "Shared Emotions." *Philosophical Explorations: An International Journal for the Philosophy of Mind and Action* 15, no. 1: 33–46.

Salmela, Mikko. 2014. "The Functions of Collective Emotions in Social Groups." In *Institutions, Emotions, and Group Agents*, edited by Anita Konzelmann Ziv and Hans Bernhard Schmid. Dordrecht: Springer, 159–176.

Sánchez Guerrero, Hector Andrés. 2016. *Feeling Together and Caring with One Another: A Contribution to the Debate on Collective Affective Intentionality.* Cham: Springer

Scheler, Max. 2008 [1923]. *The Nature of Sympathy.* New Brunswick (NJ): Transaction.

Schmid, Hans Bernhard. 2008. "Shared Feelings. Towards a Phenomenology of Collective Affective Intentionality." In *Concepts of Sharedness. Essays on Collective Intentionality*, edited by Hans Bernhard Schmid, Katinka Schulte-Ostermann, and Nikos Psarros.. Heusenstamm: Ontos, 59–86.

Schmid, Hans Bernhard. 2014. "The Feeling of Being a Group. Corporate Emotions and Collective Consciousness." In *Collective Emotions*, edited by Mikko Salmela and Christian von Scheve. Oxford: Oxford University Press, 3–22.

Schmitz, Hermann. 2003. *Was ist neue Phänomenologie?* Rostock: Koch.

Sinnerbrink, Robert. 2012. "*Stimmung*: Exploring the Aesthetics of Mood." *Screen* 53, no. 2: 148–163. https://doi.org/10.1093/screen/hjs007.

Smith, Greg M. 2003. *Film Structure and the Emotion System.* Cambridge: Cambridge University Press.

Smith, Iain Robert. 2019. "So 'Foreign' It's Good: The Cultural Politics of Accented Cult Cinema." *Continuum: Journal of Media & Cultural Studies* 33, no. 6: 705–716. https://doi.org/10.1080/10304312.2019.1677983.

Svendsen, Lars. 2005. *A Philosophy of Boredom.* London: Reaktion Books.

Thonhauser, Gerhard. 2020. "A Multifaceted Approach to Emotional Sharing." *Journal of Consciousness Studies* 27, no. 9–10: 202–227.

Von Scheve, Christian, and Sven Ismer. 2013. "Towards a Theory of Collective Emotions." *Emotion Review* 5, no. 4: 406–413. https://doi.org/10.1177/1754073913484170.

Zahavi, Dan. 2015. "You, Me, and We: The Sharing of Emotional Experiences." *Journal of Consciousness Studies* 22, no. 1–2: 84–101.

# 8 Nazi architecture as design for producing "Volksgemeinschaft"[1]

*Gernot Böhme*

## Frankfurt/M. Festhalle 1936

In his notebook on March 11, 1936, Denis de Rougemont makes the following observation:

> A murmur is going through the crowd; trumpets are to be heard from outside. Now the lightning down there in the parterre is switched off whereas light-arrows appear at the top of the big hall going down to a door at the second floor. A bright beam makes a small, brown dressed man appear at the edge—bare-headed with an ecstatic smile on his face. 40,000 people, 40,000 arms go upright by one movement. The man very slowly strides ahead greeting with a slow, bishop-like gesture and under the thunder of rhythmic "Heil-Heil-shoutings," he accepts the homage from all sides mowing step by step alongside the small path towards the rostrum. This takes six minutes—a very long period. There is nobody to notice that my hands are blocked in my pockets: they, all together, are standing upright rhythmically shouting, while staring at this very bright point, at this face with its ecstatic smile, while in the dark tears are running across their faces.
>
> Suddenly everything quiet... Him, powerful stretching out his arm, eyes to heaven—and with a hollow sound the Horst-Wessel-song goes up from the floor: 'Kam'raden, die Rotfront und Reaktion erschossen—marschier'n im Geist in unsern Reihen mit ...'
>
> Well, I got it. All of this, it's only to be grasped with a certain shiver and with your heart trembling. Now; I experience what could be called a holy tremor.
>
> I meant to take part with a mass meeting, a political demonstration. Yet, they are worshiping. Some liturgy is celebrated; the great holy ceremony of a religion is performed, of a religion which is not mine. It is overwhelming, power which pushes me back even physically, while stretching all these terribly sturdy bodies.
>
> (de Rougemont 1938, 64–67)

DOI: 10.4324/9781003131298-9

What de Rougemont tells us is, thus, you cannot comprehend what National Socialism was talking about—nothing but history, economy, and politics: "No, this is not just hatred, it's love. I was aware of love moaning from the heart of the crowd, the dump but powerful moaning of a nation, being obsessed by the man with the ecstatic smile on his face—obsessed by him, the pure and plain person, friend and savior..." So Rougemont writes the following day (66).

For Rougemont, what he sees and what he had experienced is a sort of religious emotion (Figure 8.1). Yet, he is aware of the fact that this emotion is produced. He puts down in detail how the appearance of the Führer is staged, central to the whole event 40,000 people have been waiting for the Führer over a period of four hours at the Festhalle in Frankfurt. This built-up expectation is the straw that needed nothing but a match to ignite. This expectation had been fed over days, Rougemont tells us that the drums of SS could be heard all night: two slow and three strokes following vividly. And, quite naturally the whole city was embellished and flagged. In the streets, every 100 meters, a loudspeaker. The 40,000 in the Festhalle were only a small proportion of people taking part, in fact, the whole city was included; parallel events took place in all of the 45 public spaces of the city and a huge crowd could follow by listening to a live broadcast in front of the Festhalle. Rougemont calculates that there were a million people following the event.

*Figure 8.1* May 1st. Celebration, Berlin 1936.

At the Festhalle, Hitler's performance was effectively staged: announcement by trumpets, design of light and darkness, concentration through space planning and headlights. Hitler's appearance on the second floor, moving ceremonially, interplay of greeting and homage, after thunder-like Heil-shouting, suddenly quietness. Then the Horst-Wessel-song raises, "Die Fahne hoch ..." Rougemont describes the mutual enforcement of Hitler and the crowd, what Hitler himself points to in his speech: "I am only alive if my powerful trust in the German Volk is again and again reinforced by trusting the Volk into me" (66).

## The techniques to make impression (Eindruckstechnik)

Walter Benjamin, in his essay "The Work of Art in the Age of Mechanical Reproduction" (1937) wrote about an *esthetisation of politics* (Benjamin 2008). It is true, his topics actually are the classical works of art losing their aura *and* the transformation of their cult value into an exhibition value. Behind this history of cultural damages, his essay is simultaneously the report of mass culture, the reappearance of aura and cult within the media of film, as well as in big mass-crowding with sportive events and for war. Benjamin writes: "Fascism attempts to organize the newly proletarianized masses while leaving intact the property relations which they strive to abolish. It sees its salvation in granting expression to the masses—but on no account granting them rights. The masses have a right to changed property relations; fascism seeks to give them expression in keeping these relations unchanged. The logical out-come of fascism is an aestheticising of political life" (Benjamin 2008, 41).

Yet, looking back to Benjamin's statement today we must add: it was not only that National Socialism allowed people to express their feelings, but it gave people the opportunity to have feelings and to share them. Aestheticism of politics granted not only a fake saturation, it produced real saturation of desires. The desire to be proud of something, the desire to feel one with other people, the desire of enthusiasm, and—what Rougemont rightly states—the need for love.

The German Philosopher Hermann Schmitz was much more aware of such facts than Benjamin. His book *Adolf Hitler in History* may be in danger of sympathising with National Socialism, albeit with his attempt to only understand it (Schmitz 1990). Yet, something must be learned from this book; namely, that National Socialism, in fact, gave something to people. For example, Schmitz considers what made German people become so enthusiastic about Hitler and what made people resonate with him. In response to these questions, Schmitz argues that Hitler, through his technique of producing impressions within people, actually satisfied their desires. Hitler offered an *implanting situation* that is a meaningful environment to which individual persons could feel a sense of belonging to. It is very much an art to provide these feelings—what Hermann

Schmitz calls *Eindruckstechnik*—the technique of producing certain impressions. Schmitz, with this term, denotes the same procedure which Benjamin called Aestheticism of Politics, but it encompassed more strands of application on the one hand, and it fits more explicitly to the intention to manipulate people on the other. Schmitz' concept includes modern techniques of staging, advertising, and the organisation of mass-events. This also includes what is technical in the narrower sense: loud-speakers, projectors, film, and broadcasting. Using these techniques, it was Hitler's ambition to offer the feeling of Volksgemeinschaft to the German people.

Indeed, Schmitz actually calls National Socialist staging of mass-events Plakat-Situationen. These produced the feeling of Volksgemeinschaft, but did not transform the mass into a community. Thus, similar to Benjamin's statement, Schmitz says that, via Eindruckstechnik, the feeling of Volksgemeinschaft was produced, but people were not transformed into a community.

However, aesthetic glance is a reality in itself and produces political facts. This is the reason why book titles like *Der schöne Schein des Dritten Reiches* [The brilliant glance of Third Reich] (Reichel 2006) provide underestimations of National Socialism. The very facts of Political Aesthetics of Third Reich were about eminent political issues, about mobilising the masses, and producing loyalty to the regime. It is hard to say, but the National Socialists were rather modern in this respect. They already realised the importance of media, they applied methods of marketing to the formation of political opinion, and they had strategies of public relations, in short: they introduced methods into policy making which today are designated as communication design.

Underestimating the National Socialism aesthetic strategies is a way of perpetuating suppression of its history. What is actually suppressed about it in the highest degree, what are people not willing to accept, what don't they want to remember? It is not the cruelties and crimes of National Socialism—all this has been sufficiently made public and has sufficiently been discussed. No, what people don't want to know is the fact that a large proportion of German population supported the regime, that people invested their hopes into the movement, their enthusiasm, their love, and what they called idealism at the time.

Alexander and Margarete Mitscherlich with their book on *Die Unfähigkeit zu trauern* [The inability to mourn] (Mitscherlich 1967), pointed to this fact: people were not in the position to admit that they have had positive feelings towards National Socialism and its Führer. The Mitscherlich's point this out when speaking of being in love with the Führer (Mitscherlich 1967, 71). What was impossible after 1945: to mourn for the Führer. Thus, people withdrew to the question of order and obedience—thus not admitting the individual person developed initiatives of his/her own in order to "take part with Führer's life and with his historically unique visions" (72). This was the very effect of aestheticising National Socialism policies: being included into the movement through emotional identification with the Führer.

## The architecture of uprising

The remains of National Socialism Architecture are frozen relics, in the form of structures of Zeppelin-Feld at Nuremberg; ruins of Prora; and the deconstructed Königsplatz at Munich. Picture books in which we can admire totally ruined structures like the New Reichskanzlei or planned projects like the new order of Berlin with the North-South-Traverse, all this does not really give an impression of what the architecture really was and what it intended to perform. You register some noble Neo-Classicism, some Germanic brute buildings, and some purely functional modern architecture. You do not get a feeling of the atmosphere of time, of the National Socialism movement for which this architecture constructed the frame. The architecture of National Socialism buildings and city planning was the outcome of that Eindruckstechnik, which already had been applied and had been proved before with the staging of mass-events. Hitler's architect, Albert Speer, is quite clear about this when writing his memories; he did not impress Hitler with his buildings but with his art of stage setting applied at the organisation of masse-assemblies at Tempelhofer Feld, May 1, and with the Reichspartei-Tag, 1933, at Nuremberg.

To estimate what National Socialism architecture really was, it was necessary to bodily take part with the event for which it was planned. For us, the films of Leni Riefenstahl are the best way to proceed. This is true, in particular, for the 1934 film of Reichsparteitag at Nuremberg entitled "Triumph des Willens." Generally, films are to be preferred to books seen as documents of National Socialism architecture. National Socialism in contrast to the System of Weimar Republic was understood as a movement. Ceremonies, demonstrations, and meetings were staged as critical events. With Riefenstahl's movies, we can state four characteristic patterns.

The first is the way the Führer's appearance is staged: quite naturally at the Zeppelin-Feld at Nuremberg done through symmetry and centrality, but what is also notable is the main event taking place at night. Thus, the rostrum with the Führer could be illuminated and be contrasted to the crowd. Even before that, Riefenstahl's movies show Hitler's approach to the Reichsparteitag: he comes with an airplane, and this way appearing at the sky above the crowded ancient city of Nuremberg. It's him—that is what the movie will demonstrate—he will organise the crowded lost individuals to become a Volk. It is true; the film even shows what that means: it will be military formations. The program of unity will not be an organic one but a mechanical one, as sociologists say.

The second characteristic is the emphasis on will. As well as the motto of the Reichsparteitag, Leni Riefenstahl's movie has it. In fact, the cult of the will was a main topic of National Socialism ideology. To raise the will, that was at stake. Correspondingly, the architecture of staging had the character of an appeal. With its towers, pillars, flags with long and small formats, the content of speeches and songs was architecturally set on stage: architecture as a big exclamation mark. Wake up, Germany: Deutschland erwache!

The third characteristic is that shudder which we already heard Rougemont talking about. Remembering dead people belonged to all big National Socialism marches and to its architecture: the hall of dead soldiers, the day of Schlageter, death-toll flags—by these topics and the appeal to dead heroes and martyrs, a commitment should be produced to accept their heritage and follow their example. The martyrs of the movement were already present with the 1923 Parteitag, in front of the Feldherrnhalle at Munich, represented through calls from the dark while their names were called out. The same effect was used at the Reichsparteitag at Nuremberg, staging the dedication to *Gau* and *Boden*, county and its soil: a shouting from the dark causing a deep shudder through which a primordial commitment should be felt.

The last characteristic of Riefenstahl's movies: the design of illumination. Artificial light had already been included into the spectrum of architectural means since the early 20th century, thus National Socialism here did not initiate something new. To the contrary, it is this section where the relation of National Socialism buildings to the architecture of department stores and cinemas can be shown. Yet it is true, the light domes, designed for the Reichsparteitage and then installed at the 1936 Olympic Games in Berlin, were something new and to its effect really fantastic (Figure 8.2). A space of light was produced using projectors from the anti-aircraft troops, and those domes of light horizontally defined the inner circle of the mass meeting while vertically opening it towards the endless space. That way, the known effect of Gothic domes pulling spectators upwards was intensified incredibly by feeling of *us here* assembled against a dark and hostile world outside. This feeling of a bright *us here* and a threatening, dark *other outside*, should save edification and with it the mobilisation to be ready for defence—and later for attack.

National Socialism Eindruckstechnik was used for the design of mass assemblies as well as for National Socialism buildings. This is the reason why Dieter Bartetzko is talking of National Socialism buildings as spaces set for permanent demonstrations (Bartetzko 1985, 62). It's not only the Reichsparteitag area at Nuremberg, but the Thingstätten (the Things) also, buildings like Prora, the Ordensburgen, the Gau-forums, city places and the reconstruction of cities as well as the Neue Reichskanzlei, the new chancellors building in Berlin.

As to the KdF-buildings, Prora is the most impressive one. The complex was designed as a place for holidays and recreation for more than 20,000 people. National Socialists, with areas like this and with the KdF-liners, set the beginnings of modern mass-tourism. Prora proves that during the Third Reich this was not just tourism as we know it; tourism organised for large

*Figure 8.2* Olympic Games 1936, light dome above the Olympic Stadium at Berlin.

quantities of people which nevertheless remained sort of individual vacancies for the very person. Contrary to this, Prora offers to the single person or family only very narrow, spartan facilities. Yet there were areas provided with large common spaces and forums for events (Figure 8.3). The idea was not to just provide low-income strata the opportunity to have vacancies, but the target was the Volksgenosse, the person as belonging to the Volk. This way KdF must be classified as an establishment of propaganda. KdF did not only produce posters of National Socialism charity, but was meant to provide time and space for Volksgenossen to meet as a community.

This was true more openly with buildings called Ordensburg. They were schools and training camps for party officials and elite of young men to be educated in the spirit of National Socialism. They were places of masculinity and camaraderie. Accordingly, the architecture is row and fortress like. Isolated in the mountains, they are meant to weld people together: common sleeping quarters, common sportive trainings, and common meals—this was the way Volksgenossen shared their way of life.

*Figure 8.3* Thingstätte Segeberg.

The Gau forums (first the forum at Weimar), and the National Socialism-squares—(first of all the Königsplatz at Munich), were explicitly planned as places of mass-assemblies. They were designed as stages on which the spectators were actors at the same time: horseshoe-like they contained a rostrum at one side. A main example is Munich—the capital of the movement, as it was called—where at the down end of the Königsplatz the temples of honour containing the Blutzeugen tombs got their position. Squares like this, in the history of European Urbanism are something rather new. The French revolution, and after that Hausmann, created something comparable with the field of Mars and the Place de la Concorde.

Hitler with his architect, Speer, continued with this tradition. Hitler pushed the new design of German cities. He criticised modern cities glorifying buildings of bourgeois business as the most outstanding objects, i.e., banc-houses, department stores, hotels. In his speech on culture at the 1935 Reichsparteitag, he set up ancient and medieval cities: "Ancient and medieval cities got their characteristic and admirable pattern not by big bourgeois private buildings but by the manifestations of community life" (Wolters 1941, 9). Hence, the reconstruction of German cities was not just a functional modernisation but an ideological issue, or as they said at that time an ideal one. Here again, communication design: architecture's task was to both make visible and push the experience of the German Volk as the State.

## Conclusion: the Neue Reichskanzlei

Some parting remarks on the Neue Reichskanzlei may be suitable (Figure 8.4). This building is not so much interesting as a neo-classical object but more with respect to its interior. Some authors compare the interior design with that of department stores and buildings of contemporary world exhibitions. Hitler's office, in fact, scarcely furnished, encompassed not less than 400 square meters. Yet, the claim that visitors should feel themselves little and alien at this place, at least, is single minded. It overlooks (what clearly is stated by the Mitscherlich's) that Hitler represented the personal identity for a Volksgenosse. Furthermore, through the interior of Hitler's office participated with the aura of power and he would have felt rather edified than humiliated by this environment: *Mein Führer!* This was the adequate salutation. National Socialism Eindruckstechnik this way made use what is

*Figure 8.4* Die neue Reichskanzlei: Hitler's office.

conceived of since Kant as the dialectic of the sublime: the person in danger to feel anxious and suppressed facing the over-great and over-mighty copes his ephemeral existence by feeling himself as part of something superior, something unthinkably great.

In conclusion, we can state that the representative buildings of National Socialism were—just like the spaces for marches and demonstrations— patterned by a design for mass communication. Yet, what was characteristic for the mass communication is different from what we understand by this term today, that is was communication through bodily felt presence. This design served the formation and mobilisation of the masses, understood as resurrection from humiliation and servitude. It served the establishment of the Großdeutsches Reich, the German Empire—the outcome of which, in the end, proved to be war.

## Note

1 All photos in this chapter were selected by the editor.

## References

Bartetzko, Dieter. 1985. *Illusionen in Stein: Stimmungsarchitektur im Deutschen Faschismus;ihre Vorgeschichte in Theater- und Film-Bauten*. Reinbek bei Hamburg: Rowohlt.

Benjamin, Walter. 2008. *The Work of Art in the Age of Mechanical Reproduction*. London: Penguin.

Mitscherlich, Alexander und Margarete. 1967. *Die Unfähigkeit zu Trauern. Grundlagen Kollektiven Verhaltens*. München: Piper.

Reichel, Peter. 2006. *Der schöne Schein des Dritten Reiches. Faszination und Gewalt des Faschismus*. München: Hanser.

Rougemont, Denis de. 1938. *Journal aus Deutschland 1935-1936*. Wien: Paul Zsolnay 1998, S. 64–67. Frs. Journal de l'Allemagne.

Schmitz, Hermann. 1990. *Adolf Hitler in der Geschichte*. Bonn: Bouvier.

Wolters R. 1941. *Baukunst und Nationalsozialismus: Demonstration von Macht in Europa; 1940–1943; Catalogue of the Expsiotion "Neue Deutsche Baukunst"*, Rudolf Wolters, Jörn Düwel, Niels Gutschow (eds.). Berlin: DOM Publ.

# 9 Political emotions and political atmospheres

*Lucy Osler and Thomas Szanto*

## Introduction

With the rise of Trump in America, the Brexit referendum in the UK, and the growing popularity of right-wing leaders across much of Europe, there has been talk of the increasingly nationalistic, isolationist political climate across the globe. The eruption of the Covid-19 virus, prompting international fear and blame, lockdowns, travel bans, and border closures, has arguably intensified this climate of isolation and segregation. *Political climates* are taken to be broad, mood-like framings of a country, an era, or a population, providing fertile ground for certain political ideologies and actions, while staving off others (de Rivera & Paéz 2007; Fuchs 2013a). They are typically conceptualised on the "macro" level—not applying to any particular situation or event but rather to a temporally and geographically extended time and place. We might, for example, talk of the political climate of fear that gripped Stalin's Russia. *Political atmospheres*, on the other hand, are situation specific. Using the distinction proposed by de Rivera and Paéz (2007), where climates are intended to pick out a more abstract "mood" or "tone" that characterises, say, an epoch or a nation, atmospheres emerge from concrete situations. Political atmospheres occur at the "micro" level, at specific events such as political rallies, meetings, or speeches.

While there has been a burgeoning interest in atmospheres of late, as this volume is proof of, little consideration has been paid to specifically *political* atmospheres. Yet, political atmospheres are not only a particular kind of atmosphere, they harness, create, and drive political emotions and action; they can be manipulated for political ends; and, as we will show, they can leave long shadows that impact an individual's or a community's political commitments beyond the situation in which they arise.

In this chapter, we put forward an affective, perceptual, and essentially social model of atmosphere—which we dub "interpersonal atmosphere." Interpersonal atmosphere is a relational, bodily experience that discloses the emotion, mood or affective dispositions of others present; thus, they constitute a way of experiencing others and their affective states (Osler 2021a). Using this model, we understand political atmosphere as a specific kind of

DOI: 10.4324/9781003131298-10

interpersonal atmosphere that relates to *political emotions*. We argue that political atmospheres deserve their own analysis as they not only encourage participation in shared political emotions in the moment but have a potentially long-standing impact on us due to the "sticky" character of political emotions. This leads us to raise critical questions about issues of power and responsibility regarding the emergence of political atmosphere.

In the first section, we set out our interpersonal account of atmosphere. On this account, we experience interpersonal atmospheres when we bodily grasp the emotion or mood of others, but also the various interpersonal or intragroup affective relations at play. In the following section, we highlight how the material world influences our atmospheric perception of others and can be used, and indeed manipulated, to encourage the emergence of specific emotions or moods. In section three, we highlight how we engage with atmospheres; how they move and influence our own affective states, as well as how we can impact and contribute to an interpersonal atmosphere. Having set out our interpersonal understanding of atmosphere, we turn to *political* atmospheres, arguing that political atmospheres arise from and, in turn, engender specifically *political emotions*. In section four, we set out our account of political emotions and provide a taxonomy of different types, with a particular focus on robustly shared political emotions, and in section five, bring these threads together to highlight certain unique features of political atmospheres. We explore how experiencing a political atmosphere can: crystallise our grasp of individuals as part of a political community; prompt political emotions; help integrate us around a shared affective concern with other fellow-travellers present as well as, more broadly, into a political community; and, not only affect us in the moment but can have on-going implications for what political emotions we experience and which political groups we align ourselves with.

## Interpersonal atmospheres

### Introducing atmosphere

We often talk of social situations as having an atmosphere: the festive atmosphere of a party, the raucous atmosphere of a carnival, the sombre atmosphere of a funeral, the tense atmosphere of a family argument. These are examples of what we call *interpersonal atmospheres*. We experience these atmospheres as emanating from and between people through our felt bodies.[1] Moreover, we experience them as having a particular affective tone or texture; the tense atmosphere of a family dinner feels very different to the ecstatic atmosphere of a festival. The tone of the atmosphere *tells us something* about those to whom the atmosphere relates; our sensitivity to atmosphere informs our social understanding, informs how we might act in a social situation, and being insensitive to atmosphere can lead to awkwardness, faux pas, and social missteps. In everyday experience and discourse, we are very familiar with interpersonal atmospheres. Yet, perhaps rather surprisingly, interpersonal atmospheres have received little attention.

Atmospheres, broadly speaking, have gained attention from a number of disciplines, such as architecture (e.g., Zumthor 2006; Pallasmaa 2014; Borch & Kornberger 2015), aesthetics (e.g., Dufrenne 1973; Wollheim 1993; Benjamin 2008 [1936]), management studies (e.g., Julmi 2016), psychology (e.g., Tellenbach 1968; Costa et al. 2014), geography (e.g., Anderson 2009; 2014), anthropology (e.g., Bille 2015; Bille et al. 2015; Daniels 2015), and sociology (e.g., de Rivera and Paéz 2007). Philosophical interest in atmospheres is also on the rise. Here, atmospheres have garnered a reputation for being particularly "slippery" phenomena (Böhme 1993)—something that we both feel and something that we experience as "out there" in the world, both subjective and objective. Atmospheres are described in a multitude of ways, including: as a "felt co-presence between subject and object" (Böhme 2017a, 10), as "feelings poured out spatially" (Schmitz 2019, 97), as "tuned spaces" (Binswanger 1933, 174), as "spatialized feeling" (Lipps 1935, 187), as "public moods" (Ringmar 2018), as "force fields" (Stewart 2011, 445), as "tangible, forceful, qualitative 'presences' in experiential space" (Slaby 2020, 275), as "quasi-things" (Griffero 2014, 1).

In many ways, the ambiguous ontological status of atmosphere has come to dominate the philosophical discussion of atmosphere, often being invoked as a tool to dissect, blur, and dissemble dualistic ways of thinking about subjectivity and objectivity (e.g., Griffero 2014, 2017; Böhme 2017a, b; Schmitz 2019; Slaby 2020). On dualistic conceptions, we can more-or-less clearly divide our subjective experience with what happens in "the world." This seemingly mysterious experience of a "spatialized feeling," of affectivity "out in the world," puts pressure on such dualism. Schmitz, whose name has become almost synonymous with atmosphere in the last decade (e.g., Riedel & Torvinen 2019; Nörenberg 2020; Slaby 2020), provides a discussion of atmosphere that is primarily motivated by an attempt to revolutionise the way we think about emotions and embodied subjectivity. Schmitz aims at placing emotions in public space, rather than locked in the inner sphere of an individual, and uses the notion of atmosphere to capture what he describes as a realm of pre-personal affectivity (Schmitz 2019, 97). Similarly, Griffero (2014, 2017), greatly influenced by Schmitz, uses atmospheres as an example of a new ontological category he wants to establish called "quasi-things." Slaby (2020), on the other hand, has used the notion of atmosphere to bring together the neo-phenomenology of Schmitz with Massumi's (2002, 2015) work in affect theory, moving away from an understanding of affect that centres around human bodies and human experience.

While these are all interesting, and radical, philosophical explorations, the use of atmosphere in the service of these wider projects leaves aside discussions of how experiencing atmosphere might grant us *social* understanding and rarely focus specifically upon atmospheres that arise from the expressive behaviour and interactions of people. This oversight of interpersonal atmosphere might also be attributed to the research on atmosphere being rooted in the realm of aesthetics, where atmospheres of objects and environments predominate (Böhme 2017b, 97). We propose a different starting point. In the following, we present an *interpersonal* account of atmosphere before turning to an analysis of political atmosphere. Importantly, this allows us to focus not just on what atmospheres *are* but what they *do*.

## An interpersonal approach

We suggest that we understand interpersonal atmosphere as a *way* that we experience others. We experience an interpersonal atmosphere when we grasp the expressive actions and interactions of others through our felt bodies. Atmosphere, then, is not a *what* but a *how*. When we enter a party and experience the joyful atmosphere, this is *how* we affectively experience the emotion, mood, and/or affective interconnectedness of the party. Our sensitivity to the atmosphere tells us something about those present, gives us social understanding. The *tone* of the atmosphere, as happy, conveys the affective state of the party-goers. If the affective state of the party-goers changes, so does our experience of the atmosphere. For instance, if an argument broke out at the party and all those present stopped dancing and chatting, went quiet and looked at the arguing individuals nervously, the happy atmosphere shifts. It is replaced by a tense atmosphere. As the expressive behaviour and interaction of those present changes, so does our experience of the atmosphere. The atmosphere, then, arises out of the unfolding, dynamic expressivity of those present.[2]

Influenced by Merleau-Ponty, Fuchs emphasises that we do not just visually perceive others and their expressivity experience. Rather, it is through interacting with others, as another embodied subject, that we gain interpersonal understanding (Fuchs & de Jaegher 2009; Fuchs 2013b, 2–16; Fuchs 2016). Fuchs highlights the reciprocal, affective relationship between oneself, and others as a fundamental part of our experience of others:

> Our body is affected by the other's expression, and we experience the kinetics and intensity of his emotions through our own bodily kinaesthesia and sensation. Our body schemas and feelings expand and 'incorporate' the perceived body of the other. This creates a dynamic interplay which forms the basis of social understanding and empathy.
>
> (Fuchs 2016, 198)

Fuchs' point is that our bodies are feeling bodies that resonate and are affected by the bodies of others. Experiencing others is not some affect-free, purely cognitive experience but essentially involves our felt body. Our body is the medium through which we experience the world, and that applies to our experience of other people. When interacting with someone our interaction creates interaffective "feedback cycles" (Fuchs 2020), whereby my expressions affect you and your expressions affect me and as part of the interlocking affective loop we both are attuned to, or resonate with, one another. The tense interaction between us arises not in your angry grimaces, nor my frustrated expression but in the interplay between us. What is key for our purposes is that Fuchs describes how, when individuals are engaged in such "interaffective" situations, an interpersonal *atmosphere* is created:

> This is accompanied by a holistic impression of the interaction partner and his current state (for example his anger), and by a feeling for the overall atmosphere of the shared situation (for example a tense atmosphere).
>
> (Fuchs 2016, 198)

The atmosphere, here, is not one individual's grasp of the other's discrete affective state but a bodily apprehension of the expressive interaction that arises from and between the two participants. While agreeing with much of what Fuchs says, we suggest that we can understand this bodily apprehension of others in a broader sense.[3] Audrey can, for instance, feel the atmosphere of a situation that she has only just come across, before she is an *active* participant in what is going on. For instance, Audrey may walk into a room and bodily feel the tense atmosphere of an arguing couple even if they have not noticed that Audrey has entered the room. Moreover, we suggest that this does not just apply to a dyadic interaction but to larger groups. For instance, the happy atmosphere of a party that arises out of the expressive interactions of a chatting, dancing, laughing crowd. Thus, we suggest extending Fuchs' account of interpersonal bodily perception beyond being in the middle of a dyadic interaction to a wider notion of interpersonal atmosphere.[4]

On our account, we experience interpersonal atmospheres when we bodily grasp the emotion, mood, and interconnectedness of others. Atmospheres are subjectively felt but also arise from the expressive bodies and behaviour of others out there in the world. Like all perceptual experiences (Zahavi 2003; Ratcliffe 2008), then, they are *relational*; thus, accounting for their subjective and objective character.

Three things should be noted here. First, saying that we bodily grasp the happiness of the party as an atmosphere is not to suggest that we must feel happy ourselves. Moreover, feeling an interpersonal atmosphere is not a case of emotional contagion either, where we catch the emotion of another through automatic imitation (cf. Scheler 1923 [2017]; Hatfield et al. 2011)—this would only reveal to us what we ourselves are feeling in a given situation, but would not give us the social understanding of the others. Consider how Audrey might experience the happy atmosphere of the party as accentuating her own tiredness. Her own affective state is in tension with the happy atmosphere and, as such, the atmosphere cannot be reduced to Audrey's own affective state. As we explore below, being sensitive to an interpersonal atmosphere does not require us to be swept up by it and become affectively aligned with those to whom the atmosphere relates.

Second, while we might experience the party as having a happy atmosphere that spans all those present, this is not to say that all those present must be feeling happy or performing the same expressive gesture. At the party, there are some individuals chatting away, some dancing, others listening to a funny story that is being told. There need not be a uniformity of expressivity across these individuals. Rather, the party has a certain tonal range or vitality (Stern 2010; Vendrell Ferran 2021) that spans their expressive movements. People present might be expressing themselves in different ways, undertaking different actions, but all these expressions are in the same "key" or "affective *leitmotiv*" (Walther 1922) as the others. The festivity of the party is not located in one individual, it is not just one person's happiness, rather the festivity arises from and between those present.[5] It might even be the case that there are some individuals at the party who are not happy at all, who may be sullenly sitting in the corner. However, their sullenness may not be sufficiently discordant to disrupt the dominant mood of the party.

Third, we might be more or less sensitive to an atmosphere. Indeed, in certain instances we might even get the atmosphere wrong. We might be wrong in two ways. First, we might simply get the tone of the atmosphere wrong. In the same way that one might mistake another's grimace for a smile, our experience of interpersonal atmosphere can be wrong. If Anders walks in with Audrey and remarks on the depressing atmosphere of the festive party, he has got the atmosphere wrong in an important way. This is because the atmosphere is not just a subjective feeling but reveals the emotion or mood of others and, like other forms of social cognition, is something that we can get wrong.

We might also be wrong about atmosphere in a more subtle way. Audrey might, for instance, experience the party's happy atmosphere as arising from the mood *of the group*; might take the happiness of those present to be something that they are robustly sharing as a shared emotion. Yet, it might be that those present do not experience themselves to be sharing that mood or emotion. Audrey might then experience individuals as bound together say by a common vitality or mood that they would not identify being part of. We will return to this, as it highlights how experiencing a political atmosphere can give us the potentially deceptive impression of having to do with a robustly shared political emotion.

## Material matters

One concern with an interpersonal approach is that it seems to overlook how aspects of the environment also contribute to the atmosphere of a situation. For instance, at a party, the lighting, the music, the arrangement of the furniture can all impact the affective tone of an atmosphere (Slaby et al. 2019). If our atmospheric experience is a bodily grasping of other people's expressivity, how can we account for the role that the environment plays?

We suggest that the environment can impact, shape and drive the emergence of expressive behaviour and interaction in a number of ways and, thus, plays an important role in shaping the emergence of particular interpersonal atmospheres.

i. If people are in a café crowded with furniture, their behaviour and interactions will be restricted to some extent by the environment. The occupants could not, for instance, use sweeping gestures or dance a tango in this space. There is, then, an obvious practical sense in which environmental factors shape the expressive behaviour of those present and, thus, impact what tone of atmosphere might arise.

ii. The material space of a setting might also *morph* the field of expressivity in certain ways that impact our experience of interpersonal atmosphere. In an echoey, high-ceilinged room, voices can be magnified, enhancing the sense of busy chatter that we might perceive as a bustling, dynamic atmosphere.

iii. There is, though, a more intriguing and robust sense in which the material environment can influence the emergence of an atmosphere. Not only does the world practically constrain and allow certain styles of expressive action, it also *affords* different forms of interactive possibilities. We do not perceive our environments as neutral but as permeated with certain *affordances*, solicitations for particular actions or affective possibilities (Gibson 1986; Valenti & Gold 1991; Kiverstein 2015; Hufendiek 2016; Krueger & Colombetti 2018). By affording various action, social, and affective possibilities, the environment can *solicit and drive* particular forms of interaction. The importance of this is two-fold. First, the environment can promote the emergence of particular expressive behaviour. A library might have single-person desks that are set up to afford quiet and solitary study. This encourages an individual to sit and read on their own, while discouraging library visitors forming chatty groups with one another, encouraging a quiet, still mood to emerge and, thus, giving rise to a quiet atmosphere.

Second, by affording certain styles of action and interaction, the environment can also help *sustain* particular individual and shared emotions and moods (Krueger & Colombetti 2018, 225), and thus sustain a particular atmosphere. Environmental resources can provide on-going feedback. For instance, the music at the party does not afford dancing just at the moment someone hears it but for as long as the music is playing. As such, the environment can scaffold on-going affective regulation that can help drive on-going expressive behaviour and, thus, contribute to the emergence and sustenance of a particular atmosphere.

As we highlight below, this is of particular interest in the political context; where having control of a space can *empower* one to influence the emergence of certain atmospheres, which, in turn, encourages, drives, or sustains the emergence of political emotions.

## Engaging with interpersonal atmospheres

We do not only experience interpersonal atmospheres from the "outside." We can and do experience interpersonal atmospheres as: (i) sweeping us up, (ii) as something we can participate in and contribute to, as well as (iii) change. As such, our experience of atmosphere not only arises from our grasping interpersonal expressivity and dynamics but feeds back into these very dynamics and can sustain and regulate the affective states of the participants; atmospheres, then, have a looping-effect. This will be of particular importance in the context of political atmospheres, for, as we shall argue, being exposed to political atmospheres does not just involve us being bodily sensitive to the political emotions of others but can engender, shape, and drive political emotions in us, with potentially long-standing effect.

But first, let us consider each of these ways of engaging with atmosphere in turn:

i.  When we encounter others through interpersonal atmosphere, we do not have to become like them, their expressivity does not dictate a predetermined response. We are differently sensitive to the presence of atmospheres and we become affectively involved in them in different ways (Schmitz 2019, 100). However, we can (and often do) feel swept up by interpersonal atmosphere. Having entered the party in a grumpy mood, the happy atmosphere takes hold of Audrey, leading to her feeling swept up by the happiness around her, changing her mood. Note, though, that even when we are swept up by the same atmosphere, different individuals can experience this in different ways; analogously, people experience can the same type of emotion without all experiencing it in an identical fashion (Schmid 2009; Trcka 2017).

ii.  One might not only experience being swept up by the prevailing mood of the party but also experience oneself *as part of* that atmosphere, as *contributing* to the happy atmosphere of the party. When we experience being swept up by the happy atmosphere of the party-goers, we can do so from the side-lines, enjoying the happiness sweeping over us while maintaining some detachment from the others present. Alternatively, we can join in with those present, entering into the fray. In interpersonal interactions we are not only "pre-thematically attuned to the expressivity of other" but are also "inherently concerned with how we appear to others" (Dolezal 2017, 238). As such, we not only experience our affective state aligning with those present but are aware of our own expressivity feeding into, driving and sustaining the happiness at the party. We can, then, experience ourselves as a co-producer of an atmosphere (this is perhaps closest to the description that Fuchs gives of atmosphere).

iii. It is not necessarily the case that one simply attunes and contributes to the on-going character of the atmosphere—one's presence might lead to a change in the interpersonal atmosphere. How we expressively act in relation to the others present impacts the affective tone of the atmosphere. For instance, if Audrey were to enter the party and immediately pick an argument with someone, she might be responsible for changing the mood of the party, shifting the atmosphere from a happy one to a tense one.

Some people might have more power than others to impact the tone of an atmosphere. Particularly expressive people, individuals with strong social cachet, figures of authority, and so on, all might exert a particularly strong influence over the interpersonal atmosphere that is co-produced. In turn, some people may have little sway over the behaviour of others, for instance, marginalised individuals who are ignored or not even recognised as present (Fanon 2008; Honneth 2001; Jardine 2020). We will return to this below in discussing the power and responsibility that authority figures, for example, have in the context of creating political atmosphere.

All of this is to say that experiencing interpersonal atmosphere can prompt our own affective engagement and interaction with others. However, it is important to emphasise here that the extent to which one is recognised as an expressive subject, the power that one has to impact those around one, whether one's very presence is experienced as disruptive to a social situation, all impact one's ability to engage with an atmosphere. Political, social, and normative structures, therefore, underpin one's ability to enter into certain atmospheres and enact change (Ahmed 2006; 2014). Before we turn to what we dub "political atmospheres," it must be acknowledged that our atmospheric experiences are always *shaped* by politics. In a certain sense, there are no *a*political atmospheres, to the degree that all interpersonal experience, and its material and socio-technological vehicles, is normatively and politically saturated. However, *specifically political atmospheres*, we argue, arise from and engender *specifically political emotions* and have peculiar characteristics that warrant their own analysis.

## Political emotions

We suggest that we experience a political atmosphere when we bodily grasp not just *any* emotion or mood of an individual or group but specifically *political emotions*. This has unique considerations with respect to how we engage with political atmospheres and issues of power and responsibility for shaping or encouraging the emergence of political atmospheres. To unpack the category "political atmospheres," then, we must first unpack the notion of political emotions.[6] Our discussion of political emotions will point to some distinctive characteristics of political atmospheres, in particular, their normative force, which we will detail in the final section.

## From politically focused to political emotions: a multidimensional account

Here, we outline a multi-dimensional account of political emotions. To begin with, consider some examples of paradigm cases of political emotions: a state-employee's anger in the face of the social-democratic government cutting paid lunch-breaks, which leads him to vote for the opposition; or the state employees' shared anger about this new policy, leading to nation-wide strikes; a citizen's pride in belonging to a state whose head responds resolutely, yet compassionately, to emerging white-supremacist rhetoric from the opposition; resentment, feelings of powerlessness or hatred on the part of a minority over failed retributive justice in the wake of racist police brutality; cautious hope, or its frustration, that unparalleled shifts in the socio-economic fabric of global society in the wake of the Coronavirus pandemics will have positive effects on halting climate change.

These emotions have different political stakes and impact. Moreover, they express very different social relations between the individual(s) and their relevant political communities. Roughly, we can distinguish four types of political emotion:

i. The garden-variety type of emotions in political contexts is what might be called *politically focused emotions*. The "focus" of emotions, follow-ing Helm's (2001) conceptualisation, can roughly be characterised as the aspect of emotions that indicates how the target of emotions (e.g., a given policy reform) is of a particular concern for the emoter (e.g., one's socio-economic status) and lends the target the evaluative property that the emotion tracks (e.g., the offense to one's hard-earned status).

The focus of politically focused emotions is a concern of some political import for the emoter but it is not related to any particular political community. The state-employee's anger in the face of a social-democratic government cutting paid lunch-breaks might change his vot-ing behaviour but does not involve a robust affiliation to the opposition party, any solidarity with other state-employees or the sharing of anger with fellow victims of the policy. For this type of emotion to count as political in a minimal sense it won't suffice that the target (the policy) alone is of political relevance; rather the focus must indicate some polit-ically relevant concern, a concern that has potential implications that go beyond the emoter's own, purely personal sphere and has others "in view." Audrey's repulsion of Boris Johnson as a despicable narcissist is insufficient to render her repulsion political. Even her repulsion of Johnson as a mendacious campaigner for Brexit will not suffice for it to count as a political emotion, as long as Brexit is only of her private concern, fearing, say, for her permanent residency in an EU-country. Her concern must, at least implicitly, involve Brexit as having an effect on others too.

ii. Next, we have *"socially shared political appraisal"* (Rimé 2007; cf. Michael 2011). Here, an individual's affective appraisal of a political situation is influenced and modulated by others' appraisal of that situation. Think of the different perception of one's initially covert patriotic pride over the national team's victory when witnessing the manifestation of such pride nationwide on television reports; or consider individuals' xenophobic fears reinforced by polarised peer-discussions.[7]

iii. Then we have what social psychologists call *group* or *group identification-based* political emotions. Think of feeling guilt, shame, indignation or pride "in the name of" or "on behalf" of a group that the emoter identifies with.[8] Such feelings are based on cognitive and evaluative self-categorisation as a member of a certain group, which generate evaluative and emotional identification with that group.

iv. Finally, we have what we conceive of as the fully-fledged sense of political emotions, *shared* or *collective political emotion*. To conceptualise these, we suggest applying an account to the political that has gained traction in recent debates on collective emotions, namely the collective affective intentionality account (Guerrero Sánchez 2016; 2020; see also Schmid 2009). Collective affective intentionality is the disposition of a group of individuals to *jointly* disclose situations or events in light of *shared* affective concerns.[9] Accordingly, collective *political* emotions are jointly felt appraisals of a politically relevant object, person, or event in light of the given community's political concerns. Unlike in the above categories, here we have a type of emotional sharedness of political concerns that is typically expressed linguistically by the use of the first-person plural in a robust, non-summative sense: "*We* are the 99%," "*We* are the Indignants," etc. My anger, say, is not just felt "in name of" or "on behalf of" the group I identify with and caused by something that affects that group. I'm not just angry because of the racial injustice towards my in-group. Rather, it is part and parcel of the very affective intentionality of *my* anger that it is *ours*.

But which of these types of political emotions can give rise to and, in turn, be generated, modulated or maintained by political atmospheres? We suggest that all four types can be involved in experiencing and engaging with political atmospheres, albeit with different degrees and (political) stakes. Political atmospheres may help carve out the affective focus of individuals' personal political concerns and how they relate to the others "in view" (i); the socio-material aspects of atmospheres outlined above will play a key role here. Similarly, certain political atmospheres may facilitate, shape, and frame when and how politically relevant information is socially appraised and communicatively shared among citizens (ii), while certain others, say, hostile or aggressively perceived atmospheres, will hinder such social transmission. Political atmospheres may also reinforce individuals' identification with a particular group (iii); again, socio-material aspects of atmospheres and the affordances of their dominant affective styles will chiefly factor in here. However, as robustly shared political emotions (iv) are those that most typically involve embodied interaction between individuals, we suggest that they are most likely to give rise to a political atmosphere, as well as be driven by a political atmosphere. As such, we now outline this category in more detail. In particular, we explore the *normative* power of these political emotions, as this will pave the way for unpacking the distinctive normative power of many political atmospheres.

### Robustly shared political emotions

There are four necessary and sufficient requirements for members of a community to affectively share political emotions in our robust sense:[10]

1  Two-dimensional affective intentionality requirement
2  Public recognition requirement
3  Reciprocity requirement
4  Normativity requirement

Let us briefly elaborate on these in turn:

1. First, the emotions of the members must have a *two-dimensional affective-intentional focus*: (a) a focus on the *same matter of political import*, and (b) a *background focus on the political community itself*. In other words, in collective political emotions, the sharedness of the concern and the political community itself, for which the given matter is of import, is part and parcel of the focus of those emotions. For example: For an emotion to count as a collective political emotion, it will not suffice that my repulsion of Johnson is grounded in my concern that Brexit is negatively affecting some others like myself; rather, the background focus of my emotion must also be a concern for *our political community* that is affected by Brexit in certain ways. Moreover, there is an implicit or explicit mutual *awareness* or *acknowledgement* of the sharedness of the political emotions and their import for the community. For, if one is not aware that others are sharing one's concern for our political community (and vice versa), I will not apprehend, let alone *feel*, those emotions as being shared.
2. Members of a political community must not only acknowledge the concerns as shared; they also implicitly or explicitly *claim public recognition*, i.e., recognition by third-parties, of the emotions and their import for the polity (see also Kingston 2011, 46). To be sure, often the communities in question are not in a position to *make* any public claims, because they are epistemically or politically silenced, excluded from the public sphere or might even risk negative repercussions when laying claims.[11] Nevertheless, emotions are only political if they, at least implicitly, *bear*

some *claim* to be recognised as such. The claims are typically, but *pace* Arendt (1958) not necessarily, explicitly raised in the public sphere.

This is what distinguishes *political* collective emotions from other collective emotions. Certainly, the celebration of fans at a netball match cheering for their team is also readily recognised publicly. However, there is a significant difference between sports fans and, say, a political protest group: In manifesting those concern, the latter group *makes a claim* to public recognition of their shared concerns while the former only *expresses* them without such a claim. As we will see, political atmospheres can work to magnify or silence claims for public recognition.

3.  The third requirement is that there must be certain *reciprocal* relations between the community's evaluative perspective, based on its shared concerns, and that of the members. The shared nature of political emotions must feed into the members' own affective concern for the polity, and indeed their very felt experience and emotional self-regulation. We may speak here of a certain episodic "emotional self-transformation" in episodes of robust emotional sharing. Not only are members' emotional expressions modulated by being shared but the very way in which they *experience* their own—shared—emotions. My anger over the government's policy, inasmuch as it is shared with my fellow concerned citizens, *feels* different when compared to my "private" anger. I will also have different means to *(self-)regulate* it, i.e., to influence which emotion I experience, when and how I experience and express it.[12]

    As we shall show, political atmospheres often serve to integrate the publicly shared and the personal level of political emotions, with regard to their experiential, regulatory, expressive, motivational, or agential components.

4.  The final requirement is that shared political emotions have a distinctive normativity.[13] This is arguably the most complex and most important feature for understanding shared emotions in the context of politics. The basic idea is that the sharedness of political emotions must have a certain *normative* impact on members' *emotion regulation*, their political *motivation* and *comportment* and on the *appropriateness* of their emotions. This not only places certain obligations on the emotions of members of a political community but also contributes to their integration, as well as to how individuals might feel going forward about certain matters.

    First, consider the normative function and power of political emotions vis-à-vis political concerns. It is widely agreed that emotions have certain normative functions. For instance, certain emotions testify to and voice moral breaches and aim to enforce moral or social norms. The paradigm emotions here belong to the class that Strawson (1962) famously labelled "reactive attitudes," such as resentment, contempt, indignation, moral outrage or anger, and shame (see Rawls 1971; Nussbaum 2016; Locke 2016). If we look at the normative, and political, power of political emotions it seems obvious that collective forms of resentment, for example, are more conducive to retributive justice than (inter)personal ones, such as my own resentment of a certain politician.

Second, collective affective intentionality exerts normative powers with regards individual members' emotional regulation and expression, and ultimately the very way they (ought to) feel. Consider the difference between a personal and collective commitment to feel[14] something. If I sincerely share my anger with my fellow-travellers, and am thus committed to the emotion's shared focus, it will not be appropriate to only half-heartedly join demonstrations, and even less so to ridicule in private my fellow-travellers for their naïve behaviour. If I do so, they will likely, and rightly, be sceptical of my emotional commitment to *our* cause, and might question my membership altogether. On pain of *feeling*, or in fact, *being* excluded from the community, I regulate, monitor, express, and enact my emotions accordingly. In that sense, what and how I *ought* to *feel* is partly determined by *what* and *how* we to feel *together*, and how we, as members of a community, ought to express, voice, and enact our emotions.

The normative powers of (political) emotions are typically internalised, namely by means of what Hochschild (1983) calls "feeling rules." Hochschild argues that we guide and police the appropriateness of our emotions with regard to their display, suppression, and duration by tacitly employing socioculturally inherited "emotion norms." More recently, Hochschild (2016) has introduced the concept of "deep stories." These are narratives that we are told and re-tell ourselves about how, given our political identifications and allegiances, we *ought* to feel about certain sociocultural and political issues.

Feeling rules and normatively laden emotional narratives, framed and coded by particular political communities, also apply to the perceived appropriateness of specific *types* of emotions vis-à-vis specific emotional *targets*. Telling examples are locutions such as "We are not *complainers* like those welfare-benefit recipients"; "*War-refugees* don't deserve compassion, they are traitors and should rather fight for their freedom, as *we* have"; "These are *liberal* sympathies, not *ours*" (see Hochschild 2016). As Hochschild arrestingly puts it, often "people are segregating themselves into different emotionally toned enclaves—anger here, hopefulness and trust there" (2016, 6). When such norms and internalised narratives sediment themselves abidingly, they constitute what has also been described as a particular "emotional habitus" (Illouz 2007).

Political emotions, then, are key in—normatively—*integrating* individuals' affective concerns into the broader network of the political community's concerns. They (re-)align individuals around a shared emotional perspective and enforce the political identity of groups. As such, we can conceive of experiences of robust political emotions as being "sticky." Sticky in the sense that (i) they not only impact how we *ought* to feel in a particular moment but how we *ought* to feel in similar moments going forward and (ii) they work to strengthen one's sense of belonging to a political community and reinforce one's political identity.

Finally, there is another central dimension of normativity, which is best conceived of in terms of emotions' so-called "fittingness" (D'Arms & Jacobson 2000). Whether an emotion is fitting depends not on which emotions are morally "good," or when it is wise to feel or express them or not; nor does it depend on which emotions are conducive to certain political goals. Rather, an emotion is fitting if its intentional object actually has the evaluative features that the emotion discloses to the emoter. Notice that with the notion of fittingness we need not appeal to any "objective" evaluative properties; hence, the notion doesn't depend on the endorsement of value-realism nor on any particular normative political framework (liberalism, republicanism, etc.). As such, the concept allows us to ask whether an emotion's focus picks out *those* evaluative properties that really *matter to the emoters themselves* (for more, see Szanto forthcoming).

## Political atmospheres

### Grasping a political atmosphere

What, though, does all this mean for an analysis of political atmospheres? In a simple sense, our interpersonal model of atmosphere suggests that we experience a *political* atmosphere when we bodily grasp not just any group emotion or mood but specifically political emotions and, in particular, robustly shared ones. For instance, if we attended the Christchurch massacre commemoration service, we not only bodily feel the grief of those present, we grasp the robustly shared political emotion of grief—a grief not just for those killed but a broader grief of the political community. Accordingly, the focus of this grief is not just the death of the victims but grief for this attack against religious freedom, the attack against tolerance, and equality.

We can understand this difference in terms of the demand for public recognition that this political expression of grief makes. If one stumbled across a private wake, it is likely that one would experience the wake as having a grieving atmosphere that emerges from the expressivity of those present, their tears, their hushed voices, their acts of consolation. However, what is significant about the political atmosphere of grief at the commemoration service is that the grief of those present is not merely expressed, it makes a demand on others that the grief of that political community and their shared concern for the community be recognised (and adhered to). Even when a third party, who is not participating or contributing to the political atmosphere of grief, feels this atmosphere, they can be sensitive to this demand for recognition.

Remember that this is not to say that experiencing a political atmosphere is to experience a political emotion oneself. The political atmosphere of grief at the commemoration service can be experienced by someone who has no sympathies for the cause. Indeed, they might have their own reactive political emotion to such a situation, such as feeling angry at the display of public political grief that is in conflict with one's own political community, or a reactive personal emotion, such as being annoyed that the gathering disrupts one's daily walk in Hagley Park. One can experience a political atmosphere, then, while feeling in tension with it.

Importantly, being exposed to a political atmosphere does not simply give us social understanding of the political emotion of those to whom the atmosphere relates. It can have the effect of solidifying the third party's recognition of a political community. For instance, experiencing a group of diverse individuals coming together to share a robust shared political emotion of grief with others in the wake of an attack, can work to cement the identity of that political community not just for the members of that community (as we discuss below) but also for observers. Our third party might not be swept up by or contribute to the political atmosphere of the commemoration service (perhaps staying back and observing from the side-lines) but in experiencing the political atmosphere spanning those present, it can knit together those present as a political group or community not just in that moment but going forward. The emergence of a political atmosphere, then, can have important implications for delingating, crystallising, and sustaining the identity of political communities.

This has implications in cases where certain political groups struggle to gain public recognition for their political emotions; for if such groups fail to gain public recognition, fail to have their political emotions recognised as *political* emotions and grasped in terms of *political* atmosphere, this can undermine their recognition as a political community. A recent example of this might be the activities of Black Lives Matter protesters where their anger failed to be recognised (at least in some quarters) as a political anger and led to protesters often, incorrectly, being perceived *as a mob* creating a chaotic atmosphere rather than as political activists. Getting a political atmosphere wrong, then, has broader implications than a mere social mis-understanding. Note, too, that this potentially gives third parties the power to actively undermine political identity if they wilfully refuse to recognise an emotion or an atmosphere as political, thus calling into question the relevant groups' status as a properly political community.

An inverse case should also be highlighted, namely where the experience of a political atmosphere can mislead us into thinking we are encountering a robust political community. For instance, one might perceive a political atmosphere as arising from a robustly shared emotion or collective experience, when, in fact, some or most of those involved may not experience themselves as sharing in a political emotion. Cases in point are the highly ambivalent experiences of participants in the notorious Fascist or Nazi propaganda gatherings (see Ehrenreich 2007, chap. 9). One might (as we explore below) also bodily grasp and be swept up by a political atmosphere and experience oneself as part of the political group, but you can also be wrong in this. You realise this, when others start looking at you with a quizzical expression or react hostile towards your enthusiasm. In short, we can be wrong about the scope or degree of the sharedness present (see also Szanto 2015). This reveals both the potential power and the potential problems that might emerge when we experience political atmospheres—as they can solidify the identity of a political community, even when the individual members of that community have themselves ambiguous political emotions and commitments. Political atmospheres, then, can work as powerful tools for creating the impression of robust political communities when the reality is actually more nuanced and complex.

### Being swept up by a political atmosphere

Let's now consider how we might not simply *feel* the presence of a political atmosphere but how we might *engage* with one. Like other interpersonal atmospheres, one might not just experience a political atmosphere but feel swept up by it. We suggest that we can be swept up by political atmospheres in a thin and a thick sense. One is swept up by, say, the political atmosphere of grief at the Christchurch massacre commemoration in a thin sense, when we feel that atmosphere transforming our happy mood into our own sense of grief.[15] Note, though, when we are swept up by the grief in terms of coming to feel sad ourselves, we may do so without experiencing a robust political grief together with the others.

However, we experience being swept up by that political atmosphere in a thick sense when we not only feel the atmosphere as revealing the *political* emotion of grief of the community but are swept up by that political grief in terms of feeling grief for the victims of a political attack as well as having a background concern for the political community affected by the attack. This can lead us to not only being swept up by the grief but experiencing oneself as robustly sharing in that political emotion, as part of the grieving political community.

This reveals another power that political atmospheres can have. They not only allow us to grasp the political emotions of others, not only can align our affective state with others present in a thin sense but can align and integrate our political concerns with a political community in this thicker sense. Moreover, as mentioned above, political emotions are "sticky"; they not only impact how one feels in a given moment but can affect one's identification with a political community, one's political values and norms, as well as one's future political emotions and actions. If, for instance, previously apolitical individual experiences being swept up by the political grief of the commemoration service in a thick sense, they might not only undergo a robustly shared political emotion with that group in the moment but might come to identify as a member of that political community going forward, might be motivated to join various political movements and activities, might shape their voting behaviour and cause them to undergo other political emotions (of various types) in the future. When one experiences being swept up by a political atmosphere in the thick sense, there is what we might call an on-going "hangover" effect. This is notably different to being swept up by a non-political atmosphere. For instance, being swept up by the grief of a private wake does not necessarily have this deeper impact on one's social identity or political affiliation. As we will see, it is this power of political atmospheres that make questions of manipulating, driving, and controlling such atmospheres so urgent.

## Participating in political atmospheres

As highlighted above, one might not only have one's affective state changed when we experience an interpersonal atmosphere, we can feel our own expressive actions as participating in and contributing to the atmosphere. This occurs when we not only experience our affective state aligning with those around us but engage with others in a way that our own expressivity and interactions contribute to the overarching emotion or mood.

Given the normative character of political emotions, we can understand political atmospheres as not only engendering engagement in terms of experiencing oneself swept up by that political atmosphere but as having a distinctive effect in terms of how one participates in a political atmosphere. Remember that robustly shared political emotions have a distinct normative impact on the members' emotional regulation and comportment. One not only experiences, say, political grief with the others but the way this grief is expressed and regulated is tightly structured by feeling rules.

In the third section, we outlined how interpersonal atmospheres do not *dictate* how one affectively reacts to them. Things are not so simple in the case of political atmospheres. When one participates in the political atmosphere of the commemoration, this not only involves an alignment of our affective state with others but guides and polices *how one ought to act* in that situation. As outlined above, when we share a robust political emotion, we often feel pressure to not do so "half-heartedly" but to act and express ourselves in a way that demonstrates our commitment to the shared concern. At the commemoration service, one may feel more pressure to show one's emotional engagement with the political emotion for fear of being called out for lacking sincerity or authenticity. Moreover, this is further impacted by our internalisation of "feeling rules," which guide not just how to act but indeed how to feel at public political events, and how to express one's emotions as appropriate to the concern of a political community; for example, in terms of openly expressing solidarity for the relevant political concern. Our participation, then, in a political atmosphere is more deeply and robustly normatively modulated than in non-political, interpersonal cases.

Interestingly, this might lead to a certain uniformity of expression in cases of shared political emotions as the members abide by the felt demand to engage with others and express one's commitment *in a certain way.* This can lead to a powerful political atmosphere arising as the expressivity of those present is tightly regulated and synchronised by these normative powers and, thus, the expressive tone of the group is both homogenous and intense.

Our experience as a co-producer of a political atmosphere not only has important implications for how our personal experience becomes integrated with the political community; it can also sustain the political emotion manifest in the atmosphere. Remember that when we experience ourselves as participating in an atmosphere, we are sensitive to how our own expressivity feeds into and drives the overarching emotion or mood, we recognise our own interaffective impact on *others.* This not only regulates our own commitment to and involvement with the relevant political emotion but can also work to regulate the commitment and involvement of those others, who are also sensitive to the political atmosphere at large.

As such, we can see a powerful looping effect here: where the political atmosphere not only reveals the political emotions of others, not only might sweep us up but, in turn, encourages us to act in ways that help sustain that political atmosphere, thus sustaining and regulating the shared political emotion across those present. Consequently, political atmospheres not only *arise* from political emotions but also *engender* and *sustain* them.

### Driving and influencing political atmosphere

We highlighted above a number of ways both individuals and the environment can help drive the emergence of an atmosphere, as well as sustain or change an atmosphere. Given the power that political atmospheres can have, not only in the moment but in their aftermath, it is crucial to acknowledge the responsibility and power that certain people have in manipulating the emergence of political atmosphere.

Consider again the Christchurch massacre commemoration: the New Zealand government clearly strived to create an atmosphere of shared and inclusive grief and a culture of cross-cultural understanding, compassion, and solidarity, rather than divisive fear and hatred; it did so by means of the appropriate material, discursive and symbolic scaffolds.

Let us first consider the role of Jacinda Ardern, the Prime Minister of New Zealand. As discussed above, certain individuals have particular sway over the emergence of a particular atmosphere. For instance, a particularly expressive individual at the party might, so to speak, "set the tone" of the group. People with particular social cachet or power also have disproportionate control over a group's emotions and, as such, effect on the group's atmosphere. People in positions of authority and power, for instance, might attract closer attention, their emotion or mood perceived as more salient, and, therefore, have a greater impact over the group dynamics (even those who are not directly interacting with them, may be more sensitive to that individual's expressivity over others). Empirical research into charismatic leaders suggests that leader-figures are more likely to influence the mood of a group than non-leaders (Erez et al. 2008), particularly when the leader figures are highly expressive individuals (Gooty et al. 2010; Johnson and Dipboye 2008; Sy et al. 2013).

Given this, political figures such as Ardern, have a particular responsibility when it comes to political atmospheres, due to their influence over their emergence and tone. At the commemoration service, Ardern's choice of dress (a traditional Maori cloak), the tone and content of her speech, and her expressive behaviour all worked to "set the tone" for the event. By expressively embodying the political emotion she wanted to share with her political community, and in light of the influence she has over that community as its charismatic figurehead, Ardern encouraged a particular kind of political atmosphere to emerge. Her expressivity both designated and afforded the kinds of emotions and interactions that resulted in the emergence of a political atmosphere of grief and solidarity.

Power to encourage the emergence of a political atmosphere can also be exercised in terms of manipulating the material environment in ways that are likely to afford particular political emotions in a given setting. It was not only Ardern's own affective, expressive, and behavioural style and influence which drove the political atmosphere of grief and solidarity at the commemoration. Material features were also used to help scaffold a particular political atmosphere. For instance, situating the commemoration service outside in the green fields of Hagley Park, to engender feelings of openness and accessibility to all; the performance of Cat Stevens (both in terms of his soft musical style and perhaps, symbolically, in light of him being a white man who converted to Islam); putting up photos of the victims. The scene, then, was tailored in specific ways to afford affective possibilities of coming together with others, of grieving in solidarity rather than riling up anger, of focusing on New Zealand as a multicultural state rather than instilling fear of "outsiders," of highlighting tolerance and sympathy not aggression and hostility.

Indeed, consider how a different material setting could have afforded the emergence of a very different political atmosphere; say if the government had pinned up posters of the attacker with slogans demanding for retribution, if people had not been able to occupy one shared space but been forced to sit or stand in different areas, if angry music had been played, and so on. What this emphasises is that having control over material environments can also give one power in terms of encouraging certain political atmospheres (and thus emotions) to emerge.

If one comes from a grassroots political community, with little access to material space, this way of influencing the emergence of a political atmosphere is not as easily available to you—you must "make do" with the environments you have, rather than tailor them to your own ends. This can add to the obstacles that certain political communities face in gaining public recognition for their political emotions and atmospheres, as well as their political community more generally.

Importantly, this suggests that we are not all equal when it comes to political atmospheres—our actions and expressivity do not have equal weight when it comes to encouraging the emergence of a political atmosphere and their political emotions. This should be of particular concern when we think about how figureheads might use their power to manipulate the emergence of certain atmospheres that will benefit their political community over others, or indeed benefit their own personal political agenda within a political community.

## Conclusion

In this chapter, we have presented an interpersonal model of atmosphere and used this to explore specifically political atmospheres. As political atmospheres arise from and engender expressions of political emotions (in particular, robustly shared ones), we suggest that political atmospheres have various unique features. Grasping a political atmosphere can crystallise our recognition of political groups; being swept up by a political atmosphere can align not only our affective state with the group but our political identity; experiencing oneself as a co-producer of a political atmosphere can further integrate our own and others' experience of sharing a robust political emotion; having power (in terms of having control over the social and material environment of a situation) can allow one to manipulate and shape the emergence of a political atmosphere. Political atmospheres, then, are powerful tools for both driving political emotions and for forging and sedimenting political communities and political identities. Perhaps, though, what is most interesting about political atmospheres and the political emotions they involve is that they not only affect us in the moment but can shape our on-going political engagements, political affiliations, and political actions.

It is worth noting that in the above, we have presupposed that our experience of an interpersonal political atmosphere occurs when we are in the same physical spaces as the relevant others. However, many of our political interactions take place in the online sphere (Papacharissi 2015). Indeed, for many of us this might be our main arena for engaging with political communities, political events and political expression. By adopting an *interpersonal* model of political atmosphere, whereby we understand our experience of political atmosphere in terms of our bodily grasping (and engagement) with the political emotions of those present as atmosphere, there might be concern that our approach sheds little to no light on political emotions or atmospheres in the online world. However, this concern, we suggest, is premised on a conception of our experience of others online being disembodied and our online interactions as being unable to be interaffective. We suggest, contrary to this conception, that we can and do experience others' expressivity online and that we can and do experience one another interaffectively online (Osler 2020, 2021b; Kekki 2020; Osler & Krueger 2021). Thus, we take it that our account of political atmospheres can be readily applied to the online sphere. The affective nature and political force of atmospheres will not evaporate or become merely "virtual," when people share their political concerns online. Quite the contrary, atmospheres underpinning and shaping online political engagements can be just as "real" as those experienced in physical co-presence, and given their virtually infinite scope may even be more powerful.

## Acknowledgements

Work on this chapter was generously supported by an Austrian Science Fund (FWF) grant for the research project "Antagonistic Political Emotions" (P 32392-G) as well as, the Carlsberg Foundation project "Who are We? Self-identity, Social Cognition, and Collective Intentionality" (CF18-1107).

## Notes

1 One of the few things that atmosphere researchers agree on is that there is no such thing as an unfelt atmosphere (Fuchs 2013a; Griffero 2014, 2017; Böhme 2017a, b; Slaby 2020).
2 Note that this marks a crucial difference between our account and Schmitz's conception of atmospheres as a pre-personal realm of affectivity. While Schmitz conceives of emotions as atmospheres out in public space, it is not clear how atmospheres relate to the world more generally. As Böhme (2017a, 17) puts it: "Schmitz's approach suffers above all from the fact that he credits atmospheres with too great an independence from things. They float free like gods and have as such nothing to do with things, let alone being their product"—and we would add to this, too great an independence from the people who they relate to.
3 Indeed, we would go so far to defend the idea that we can bodily apprehend others even in the case of mediated social experiences online and, thus, can experience interpersonal atmospheres online. See: Krueger & Osler 2019; Osler 2020, 2021b; Kekki 2020; Osler & Krueger 2021 for work on online sociality that would underpin such an argument.
4 For a more detailed account of our experience of interpersonal atmosphere as a fully-embodied form of empathy, see Osler (2021a).
5 Though this is not to say that we experience the atmosphere as equally distributed across those present. There may, for instance, be a concentration of atmosphere around someone particularly expressive (Trigg 2020; Osler 2021a).
6 For a more detailed account of political emotions, which we have adapted and revised here, see Szanto & Slaby 2020.
7 Such social appraisal mechanisms are partly responsible for so-called "emotional enclaves" (Hochschild 2016), especially if they are amplified by social or mass media.
8 Note that the group with which one identifies might turn out to be imaginary; see Szanto & Montes Sánchez Forthcoming.
9 Notice that the account doesn't require any emergent or supra-individual subject as the bearer of that disposition; rather it is actualised by individual members' ability to "feel-towards together" (see Guerrero Sánchez 2016).

10  For alternative, but similar, requirements for not necessarily political but collective emotions in general, see Szanto (2015, 2018); León et al. (2019).
11  Thanks to Michaela Mihai for pressing us on this point.
12  See more on (disruptions of) social and collaborative emotion regulation in Szanto 2017a.
13  Note that ordinary individual emotions are also subject to and guided by what sociologists of emotions call "feeling rules" or "emotion norms" (Hochschild 1983; see Szanto & Slaby 2020). But the way and grade in which normativity permeates shared emotions are distinctive of these. For related differences regarding the normativity of individual and collective imagination see Szanto 2017b and Szanto & Montes Sánchez 2021)
14  When we talk about "collective commitment to feel," a word of caution is in order: we do not endorse here the prominent but much-criticised problematic, "joint commitment" account of collective emotions of Gilbert (e.g., 2002, 2014), but only use the notion to highlight *one* specifically *normative* difference between individual and collective emotions—in contrast to Gilbert who takes it to be in fact definitive of collective emotions.
15  See Schmitz (2019) for an interesting discussion on the "authority" that certain emotions have over others.

# References

Ahmed, Sara. 2006. *Queer Phenomenology: Orientations, Objects, Others.* Durham and London: Duke University Press.
Ahmed, Sara. 2014. *The Cultural Politics of Emotion.* Edinburgh: Edinburgh University Press.
Anderson, Ben. 2009. "Affective Atmospheres." *Emotion, Space and Society, 2*(2): 77–81.
Anderson, Ben. 2014. *Encountering Affect: Capacities, Apparatuses, Conditions.* London: Ashgate.
Arendt, Hannah. 1958. *The Human Condition.* Chicago: University of Chicago Press.
Benjamin, Walter. 2008. *The Work of Art in the Age of its Technological Reproducibility, and Other Writings on Media.* Cambridge, MA: Harvard University Press.
Bille, Mikkel. 2015. "Lighting Up Cosy Atmospheres in Denmark." *Emotion, Space and Society, 15*: 56–63.
Bille, Mikkel, Peter Bjerregaard, and Tim Flohr Sørensen. 2015. "Staging Atmospheres: Materiality, Culture, and the Texture of the In-Between." *Emotion, Space and Society, 15*: 31–38.
Binswanger, Ludwig. 1933. "Das Raumproblem in der Psychopathologie." *Ausgewählte Vorträge und Aufsätze, 2*: 174–225.
Böhme, Gernot. 1993. "Atmosphere as the Fundamental Concept of a New Aesthetics." *Thesis eleven, 36*(1): 113–126.
Böhme, Gernot. 2017a. *The Aesthetics of Atmospheres.* London, New York: Routledge.
Böhme, Gernot. 2017b. *Atmospheric Architectures: The Aesthetics of Felt Spaces.* London: Bloomsbury.
Borch, Christian, and Martin Kornberger. ed. 2015. *Urban Commons: Rethinking the City.* London, New York: Routledge.
Costa, Cristina, Sergio Carmenates, Luis Madeira, and Giovanni Stanghellini. 2014. "Phenomenology of Atmospheres. The Felt Meanings of Clinical Encounters." *Journal of Psychopathology, 20*: 351–357.

Daniels, Inge. 2015. "Feeling at Home in Contemporary Japan: Space, Atmosphere and Intimacy." *Emotion, Space and Society*, *15*: 47–55.

D'Arms, Justin, and Daniel Jacobson. 2000. "Sentiment and Value". *Ethics*, *110*(4): 722–748.Dolezal, Luna. 2017. "The Phenomenology of Self-Presentation: Describing the Structures of Intercorporeality with Erving Goffman." *Phenomenology and the Cognitive Sciences*, *16*(2): 237–254.

Dufrenne, Mikel. 1973. *The Phenomenology of Aesthetic Experience*. Evanston: Northwestern University Press.

Ehrenreich, Eric. 2007. *The Nazi Ancestral Proof: Genealogy, Racial Science, and the Final Solution*. Bloomington: Indiana University Press.

Erez, Amir, Vilmos F. Misangyi, Diane E. Johnson, Marcie A LePine, and Kent C. Halverson. 2008. "Stirring the Hearts of Followers: Charismatic Leadership as the Transferal of Affect." *Journal of Applied Psychology*, *93*(3): 602–615.

Fanon, Frantz. 2008. *Black Skin, White Masks*. Translated by Richard Philcox. New York: Grove Press.

Ferran, Íngrid Vendrell. 2021. "How to Understand Feelings of Vitality: An Approach to Their Nature, Varieties, and Functions." In *Phenomenology of Bioethics: Technoethics and Lived-Experience* edited by Susi Ferrarello, 115–130. London: Springer.

Fuchs, Thomas. 2013a. "The Phenomenology of Affectivity." In *The Oxford Handbook of Philosophy and Psychiatry* edited by Kenneth Fullford et al., 612–631. Oxford: Oxford University Press.

Fuchs, Thomas. 2013b. "Depression, Intercorporeality, and Interaffectivity." *Journal of Consciousness Studies*, *20*(7–8): 219–238.

Fuchs, Thomas. 2016. "Intercorporeality and Interaffectivity." In *Intercorporeality: Emerging Socialities in Interaction*, edited by Christian Meyer, Jürgen Streek, and J. Scott Jordan, 194–209. Oxford: Oxford University Press.

Fuchs, Thomas. 2020. "Embodied Interaffectivity and Psychopathology." In *The Routledge Handbook of Phenomenology of Emotion*, edited by Thomas Szanto and Hilge Landweer, 323–336. London, New York: Routledge.

Fuchs, Thomas, and Hanne De Jaegher. 2009. "Enactive Intersubjectivity: Participatory Sense-Making and Mutual Incorporation." *Phenomenology and the Cognitive Sciences*, *8*(4): 465–486.

Gibson, James. 1986. *The Ecological Approach to Visual Perception*. Hillsdale: Lawrence Erlbaum Associates.

Gilbert, Margaret. 2002. "Collective Guilt and Collective Guilt Feelings." *The Journal of Ethics*, *6*(2): 115–143.

Gilbert, Margaret. 2014. "How We Feel: Understanding Everyday Collective Emotion Ascription." In *Collective Emotions. Perspectives from Psychology, Philosophy, and Sociology* edited by Christian von Scheve, and Mikko Salmela, 17–31. Oxford: Oxford University Press.

Gooty, Janaki, Shane Connelly, Jennifer Griffith, and Alka Gupta. 2010. "Leadership, Affect and Emotions: A State of the Science Review." *The Leadership Quarterly*, *21*(6): 979–1004.

Griffero, Tonino. 2014. *Atmospheres: Aesthetics of Emotional Spaces*. Farnam: Ashgate.

Griffero, Tonino. 2017. *Quasi-Things: The Paradigm of Atmospheres*. New York: Suny Press.

Guerrero Sánchez, Héctor Andrés. 2016. *Feeling Together and Caring with One Another: A Contribution to the Debate on Collective Affective Intentionality*. Cham: Springer.

Hatfield, Elaine, Richard L. Rapson, and Yen-Chi Le. 2011. "Emotional Contagion and Empathy." In *The Social Neuroscience of Empathy* edited by Jean Decety and William Ickes, 19–30. Cambridge: MIT Press.

Helm, Bennett W. 2001. *Emotional Reason. Deliberation, Motivation, and the Nature of Value*. Cambridge: Cambridge University Press.

Hochschild, Arlie Russell. 1983. *The Managed Heart. Commercialization of Human Feeling*. Berkley: University of California Press.

Hochschild, Arlie Russell. 2016. *Strangers in their Own Land. Anger and Mourning on the American Right*. New York: The Free Press.

Honneth, Axel. 2001. "Invisibility: On the Epistemology of 'Recognition'." *Aristotelian Society Supplementary Volume*, 75(1): 111–126.

Hufendiek, Rebekka. 2016. "Affordances and the Normativity of Emotions." *Synthese*, 194(11): 4455–4476.

Illouz, Eva. 2007. *Cold Intimacies. The Making of Emotional Capitalism*. Cambridge: Polity.

Jardine, James. 2020. "Social Invisibility and Emotional Blindness." In *Perception and the Inhuman Gaze: Perspectives from Philosophy, Phenomenology, and the Sciences* edited by Anya Daly, Fred Cummins, James Jardine and Dermot Moran. London, New York: Routledge.

Johnson, Stefanie K, and Robert L. Dipboye. 2008. "Effects of Task Charisma Conduciveness on the Effectiveness of Charismatic Leadership." *Group & Organization Management*, 33: 77–106.

Julmi, Christian. 2016. "The Concept of Atmosphere in Management and Organization Studies." *Organizational Aesthetics*, 6(1): 4–30.

Kekki, Minna-Kerttu. 2020. "Authentic Encountering of Others and Learning through Media-Based Public Discussion: A Phenomenological Analysis". *Journal of Philosophy of Education*, 54(3): 507–520.

Kingston, Rebecca. 2011. *Public Passion: Rethinking the Grounds for Political Justice*. Montreal: McGill-Queen's Press.

Kiverstein, Julian. 2015. "Empathy and the Responsiveness to Social Affordances." *Consciousness and Cognition*, 36: 532–542.

Krueger, Joel, and Giovanna Colombetti. 2018. "Affective Affordances and Psychopathology." *Discipline Filosofiche*, 28(2): 221–246.

Krueger, Joel, and Lucy Osler. 2019. "Engineering Affect: Emotion Regulation, the Internet, and the Techno-Social Niche." *Philosophical Topics*, 47(2): 1–57.

León, Felipe, Thomas Szanto, and Dan Zahavi. 2019. "Emotional Sharing and the Extended Mind." *Synthese*, 196(12): 4847–4867.

Locke, Jill. (2016). *Democracy and the Death of Shame: Political Equality and Social Disturbance*. Cambridge: Cambridge University Press.

Lipps, Theodor. 1935. *Empathy, Inner Imitation, and Sense-Feelings. A Modern Book of Aesthetics*. New York: Holt and Company.

Massumi, Brian. 2002. *Parables for the Virtual: Movement, Affect, Sensation*. Durham: Duke University Press.

Massumi, Brian. 2015. *Politics of Affect*. Cambridge: Polity.

Michael, John. 2011. "Shared Emotions and Joint Action." *Review of Philosophy and Psychology*, 2(2): 355–373.

Nörenberg, Henning. 2020. "Hermann Schmitz." In *The Routledge Handbook of Phenomenology of Emotion* edited by Thomas Szanto and Hilge Landweer, 215–223. London, New York: Routledge.

Nussbaum, Martha C. 2016. *Anger and Forgiveness: Resentment, Generosity, Justice.* Oxford: Oxford University Press.

Osler, Lucy. 2021a. *"Interpersonal Atmospheres: An Empathetic Account."* PhD thesis, University of Exeter, Exeter.

Osler, Lucy. 2021b. "Taking Empathy Online." *Inquiry*, 1–28.

Osler, Lucy. 2020. "Feeling Togetherness Online: a Phenomenological Sketch of Online Communal Experiences." *Phenomenology and the Cognitive Sciences*, *19*(3): 569–588.

Osler, Lucy, and Joel Krueger. 2021. "Taking Watsuji Online: Betweenness and Expression in Online Spaces." *Continental Philosophy Review*. DOI: 10.1007/s11007-021-09548-7.

Pallasmaa, Juhani. 2014. "Space, Place and Atmosphere. Emotion and Peripherical Perception in Architectural Experience." *Lebenswelt. Aesthetics and Philosophy of Experience*, 4: 230–245.

Papacharissi, Zizi. 2015. *Affective Publics: Sentiment, Technology, and Politics.* Oxford: Oxford University Press.

Ratcliffe, Matthew. 2008. *Feelings of Being: Phenomenology, Psychiatry and the Sense of Reality.* Oxford: Oxford University Press.

Rawls, John. 1971. *A Theory of Justice.* Cambridge, MA: Harvard University Press.

Riedel, Friedlind, and Juha Torvinen. eds. 2019. *Music as Atmosphere: Collective Feelings and Affective Sounds.* London, New York: Routledge.

Rimé, Bernard. 2007. "The Social Sharing of Emotion as an Interface between Individual and Collective Processes in the Construction of Emotional Climates." *Journal of Social Issues*, *63*(2): 307–322.

Ringmar, Erik. 2018. "What are public moods?" *European Journal of Social Theory*, *21*(4): 453–469.

Rivera, Joseph de, and Dario Páez. 2007. "Emotional Climate, Human Security, and Cultures of Peace." *Journal of Social Issues*, *63*(2): 233–253.

Scheler, Max. 1923 [2017]. *The Nature of Sympathy.* Translated by P. Heath. London: Routledge.

Schmid, Hans Bernhard. 2009. *Plural Action: Essays in Philosophy and Social Science* (Vol. 58). New York: Springer Science & Business Media.

Schmitz, Hermann. 2019. *New Phenomenology: A Brief Introduction.* Milan: Mimesis.

Slaby, Jan. 2020. "Atmospheres: Schmitz, Massumi and Beyond." In *Music as Atmosphere* edited by Friedlind Riedel and Juha Torvinen, 274–285. London, New York: Routledge.

Slaby, Jan, Rainer Mühlhoff, and Philipp Wüschner. 2019. "Affective Arrangements." *Emotion Review*, *11*(1): 3–12.

Stern, Daniel N. 2010. *Forms of Vitality: Exploring Dynamic Experience in Psychology, the Arts, Psychotherapy, and Development.* Oxford: Oxford University Press.

Stewart, Kathleen. 2011. "Atmospheric Attunements." *Environment and Planning D: Society and Space*, *29*(3): 445–453.

Strawson, Peter F. 1962. "Freedom and Resentment." *Proceedings of the British Academy* 48, 187–211.

Sy, Thomas, Jin Nam Choi, and Stefanie K. Johnson. 2013. "Reciprocal Interactions Between Group Perceptions of Leader Charisma and Group Mood through Mood Contagion." *The Leadership Quarterly*, *24*(4): 463–476.

Szanto, Thomas. forthcoming. "Can it Be or Feel Right to Hate? On the Appropriateness and Fittingness of Hatred." *Philosophy and Society*.

Szanto, Thomas. 2018. "The Phenomenology of Shared Emotions: Reassessing Gerda Walther." In *Women Phenomenologists on Social Ontology. We-Experiences, Communal Life, and Joint Action*. Edited by Sebastian Luft, and Ruth Hagengruber, 85–104. Dordrecht: Springer.

Szanto, Thomas. 2017a. "Collaborative Irrationality, Akrasia and Groupthink: Social Disruptions of Emotion Regulation." *Frontiers in Psychology* 7(2002): 1–17.

Szanto, Thomas. 2017b. "Collective Imagination: A Normative Account." In *Imagination and Social Perspectives: Approaches from Phenomenology and Psychopathology* edited by Michela Summa, Thomas Fuchs, and Luca Vanzago, 223–245. London, New York: Routledge.

Szanto, Thomas. 2015. "Collective Emotions, Normativity, and Empathy: A Steinian Account." *Human Studies*, *38*(4): 503–527.

Szanto, Thomas, and Alba Montes Sánchez. forthcoming: "Imaginary Communities, Normativity and Recognition: A New Look at Social Imaginaries." *Phänomenologische Forschungen*.

Szanto, Thomas, and Jan Slaby. 2020. "Political Emotions." In *The Routledge Handbook of Phenomenology of Emotions* edited by Thomas Szanto and Hilge Landweer. London, New York: Routledge, 478–492.

Tellenbach, Hubertus. 1968. *Geschmack und Atmosphäre*. Salzburg: Otto Müller.

Trcka, Nina. 2017. "Collective Moods. A Contribution to the Phenomenology and Interpersonality of Shared Affectivity." *Philosophia*, *45*(4): 1647–1662.

Trigg, Dylan. 2020. "The Role of Atmosphere in Shared Emotion." *Emotion, Space and Society*, 35: 100658. https://doi.org/10.1016/j.emospa.2020.100658

Valenti, S. Stravos, and James MM. Gold. 1991. "Social Affordances and Interaction I: Introduction." *Ecological Psychology*, *3*(2): 77–98.

Walther, Gerda. 1922. "Zur Ontologie der sozialen Gemeinschaften." *Jahrbuch für Philosophie und phänomenologische Forschung*.

Wollheim, Richard. 1993. *The Mind and Its Depths*. Cambridge, MA: Harvard University Press.

Zahavi, Dan. 2003. "Intentionality and Phenomenality: A Phenomenological Take on the Hard Problem." *Canadian Journal of Philosophy*, *33*(1): 63–92.

Zumthor, Peter. 2006. *Atmospheres: Architectural Environments, Surrounding Objects*. Berlin: Birkhäuser.

# Conclusion
## Something we all share

*Andreas Philippopoulos-Mihalopoulos*

### Meta-atmosphere

What is shared in an atmosphere? Is it emotions? Sensorial stimulation? Corporeal presence? A feeling of belonging?

Atmospheric literature mostly answers with a sonorous "yes" to all the above (please forgive the violence of grouping all literature together, but I am trying something here), encouraging the observation of the space between bodies and things, finding smoothness in striation, and even self-assuredly allowing for variations, gradations, and deviations from these emotional gatherings of thought. From its German theoretical beginnings to its Scandinavian evolution, with the grand admixture of Italian, French, British, Australian, and a few other (but not many other) schools of thought or at least approaches, the current atmospheric literature pendles in that ineffable space between the human and the spatial.

That dwelling space of atmosphere is apparently difficult to capture. Atmospheres are elusive, says the literature. So the literature brings in the affective (Anderson 2009), the sensorial (Bille *et al.* 2015), the habitual (Ahmed 2006), and other such tools with which to adumbrate this notoriously elusive thing.

There is a lot of sharing within that space, whether through imitation, repetition, diffusion, affordance, attunement, transmission, or outright contagion. Sharing takes place on a choice of levels such as inter- or super-subjective, intra-active, inter-agential or at least between subject and object, preconscious or precognitive, and regularly half-a-second before the emergence of the atmosphere in question (Thrift 2007). These are all ways in which we grapple with atmospherics, our toolbox that increasingly becomes sharper and clearer in its cleaving categorisations of who is who and what is what when it comes to this most evanescent of things.

So far, nothing new. Because of course this has always been our method. We ("humans") posit a divine we do not understand, and we subsequently employ hotly divergent categories of religious rituals in order to envelop that divine in a securely protective layer of perpetual tenuousness. What I mean is this: we (atmospherologists) have posited an evanescence we do

DOI: 10.4324/9781003131298-12

not understand ("god or atmosphere," to paraphrase Spinoza's *Deus sive Natura*) and are currently in the process of developing an extremely sophisticated register of methods, concepts, tools, and approaches to "capture" atmospheres. But kindly note the oddity: all these tools are characterised by a paradoxical and slightly duplicitous doubling-up: in name, they help us clarify the atmospheric; in substance, they ensure that atmospheres remain as elusive as ever.

For if we are enamoured with atmospheres, it is in a deeply Lacanian way. Atmosphere is our *petit objet a*. We wouldn't know what to do with it if it ever reciprocated our love and decided to appear fully before us.

There is a simple meta-explanation for all this. Atmospheric literature is itself an atmosphere. It loves to bathe in the pathos of the romanticised indefinable, the *je ne sais quoi* of the uncanny, the imperceptibility of belonging, the fuzziness of *hygge* cosiness. And why not! These are attractive things, nurtured by centuries-long sedimented nostalgia canonised within a phenomenological tradition nobly anthropocentric, solidly consciousness-based (even when consciousness is skilfully shown to be momentarily bypassed), richly ennobled by architectural theory, convincingly situated alongside anthropology, ethnography, sociology, and crowned by stellar philosophical credentials.

I am not being sarcastic. Or perhaps ever so slightly. The science is all solid and good. The political project, however, is problematic (bar notable exceptions): at best unexamined, and at worst attempting to be objective and, therefore, ending up fully complicit.

But we are forgiven if we forget this. "How lovely it is that we forget," sings Zarathustra (Nietzsche 2005). We are all part of this atmosphere too, remember? This is the atmosphere of atmospherologists, the ones who write in this volume, the ones who read this volume. This atmosphere is as rarefied as it is welcoming. It includes historical lessons, future tendencies, and a widely spanning interdisciplinarity: any disciplinary approach can deal (and most indeed have dealt) with atmospherics and contribute to the theoretical and empirical development of atmospheres. It is holistic in that it leaves little outside. But holism goes the other way too: this literary atmosphere includes ("engulfs" as I show below) even political tensions and deviations from the theory itself. Any objection to atmospheres is well ensconced in the emotional charge that comes with atmospheres. Incredulity and fear, suspicion and doubt, resistance and outright attack against this all-cannibalising luminous divinity that we have named atmosphere: all this fits snugly under the umbrella of atmospheric adulation.

For can we, honestly, hand in heart, we who are critical of atmospheres, we who attack their nefarious political smoke and mirrors game, we who warn others of how atmospheres manipulate us and how we are not in control, we amongst the atmospherologists who gleefully point out the naivete of the ones who simply observe, classify, experience, assess, or simply (simply!) theoretically immerse themselves into atmospheres; can we really say that we are not as besotted or more with atmospheres as all the blithe ones?

I have now set up our atmosphere. We are all enveloped in its allure. We all share the emotion.

Let's go deeper.

## All around us

We share something else too, something all around us. We share a single planetary atmosphere. This is not just a question of the weather (whose effect on emotions is well-documented: Griffero 2014) but indeed the original, non-metaphorical iteration of the atmospheric word and concept. Planetary atmosphere, and by extension the kind of atmosphere we are talking about here, is a totality that, not unlike seduction, is characterised by a force of attraction that does not easily let bodies escape. Atmosphere is an enclosure, a *sphere* of *air* and *mist* as the etymology shows, whose protective services make it difficult for anyone either to break away from it, or indeed to enter it. Atmosphere attracts and destroys at the same time.

While this admittedly banal reference to the earth's atmosphere is mostly omitted from the kind of atmospherics discussed in this volume, I have good reasons to include it: first, I am trying to posit here a posthuman understanding of atmospheres that goes beyond the distinction between human and nonhuman and, by extension, subject and object. A return to the original planetary dimension of the concept of atmosphere seems a good basis for this. Second, I am hoping to move the locus of emotions from the human to the nonhuman—a little like the way we all connect to the meteorological—and observe how atmospherics deal with this continuum between the various bodies. Third, I would like to play with this illusion: that the planetary atmosphere is *all around* us. It is not. In the era of the Anthropocene, the strife to move away from the human is becoming harder. The continuum between humanity and atmosphere is not only established but indeed manipulated so that we are brought at the point where the atmosphere needs salvaging from us and our actions. So, this atmosphere at least (and as we shall see, all atmospheres, meteorological or not) emerge *through* us. So neither just around us nor, however, just through humanity. But we like to think that way. It is comforting. We, humans, are not the same as animals, plants, or objects after all, are we. And we place that difference away from us, at the end of our skins, floating in the aether. Another atmosphere is at work here.

In this text, I am playing with the chiaroscuro of atmospherics. I posit three paradoxes or at least turbulent dualities that are at play when it comes to atmospherics.

The first one is the paradox of the need for atmospheres on the one hand, and the need to withdraw from them on the other. They are both ontological needs, tightly woven with the texture of atmosphere itself. Atmospheres are needed as safety, community, belonging, return, understanding, agreement.

A withdrawal from atmospheres is needed to combat false constructions of desire, problematic exclusions, invisible exploitations, ethical compromises. These two needs must co-exist and the only way to do this is through our continuous movement from one to the other.

The second paradox is the generally agreed (and no less problematic for this reason) need for atmospheres to be perceived as elusive. This is a need pertaining to the atmospheric sphere as it were, in which we, the ones thinking and writing about atmospheres, partake. The ontology of atmospherics envelops (to use Derek McCormack's 2018 term) the way atmospheres are to be approached phenomenologically. It is, however, in the interest of atmospheres to carry on with this impression of elusiveness, because this is the only way in which they can remain persuasive. But as I show in the last section of this text, there is nothing ineffable about atmospheres.

The third paradox I will be juggling here is the role of the human. While seemingly in the centre of it all, and especially when we think of atmospheric constructions in the Anthropocene, the human is actually expelled from the core of atmospheric construction, leaving behind only one element: desire. Atmospheres seemingly include the human as co-creator or at least participant, but, in fact, they are dehumanising machines that reduce the human to a circular desiring machine. Still, the *mediated* human, namely the post-Enlightenment, collective, posthuman human animal, is our only hope for dealing with the atmospheric issue.

The text, therefore, is structured as such: in what follows, I expand the current thought on the locus of emotion as outside the human, and argue that emotions not only are produced "externally" (in reality, such distinction is not valid since interior and exterior are in a continuum with only artificial divisions) but actually are *of* nonhuman entities. In that sense, emotions like percepts, are planetary rather than just human. This means that sharing emotions is sharing a different materiality that cannot be adequately described by the usual phenomenological mechanisms. In these "neo-Homeric" times as I rather playfully call our time, sharing becomes an ethical question and tool towards a reimagining of different atmospheric constructions. By that point, however, it becomes clear that I am cautioning against atmospherics that create conditions of immunity, safety, belonging etc., since they all rely on exclusion and indeed exploitative engulfment of otherness (human and nonhuman). In section five, I offer a summary of the four most important conditions needed to be met in order for an atmosphere to be successfully engineered—but I offer these as *causes* that we can explore when we try to withdraw from atmospheric attraction, causes behind our perceived desires for more atmospherics. Finally, in section six, I suggest withdrawal from atmospherics but in a climate of understanding of both the need to remain in a protective atmosphere and the need to rupture it and look for this truly elusive outside (as opposed to the falsely elusive atmospheric inside).

## A storm is brewing

A meteor storm, that is. Although the occasional meteor shrapnel does hit the earth, everyone knows that these storms are more spectacular than lethal. The same thing that attracts them is luckily the one that burns them. Our planet is the Homeric Island of the Sirens. Our atmospheric song attracts, our atmospheric fangs devour.

Atmosphere would not exist without gravity. Gravity, however, is not an a priori but a co-emergent of the atmosphere. Gravity waves, for example, originate in atmospheric events (such as mountain top winds or jet streams) and affect the middle and outer atmosphere, keeping things moving (Hines 1972). Gravity needs the atmosphere in order to remain animated, just as atmosphere needs gravity in order to stay put. This combination of pause and movement, stability and speed, is at the core of the atmospheric paradox. Atmosphere attracts because of its pulsating velocity, its continuous yet imperceptible movement, a static yet vertiginous movement as the second etymology of the word shows, where *sphere* (in modern Greek σφαίρα, "sfaira") stands for *missile* or *bullet*.

Our atmosphere is constituted of various chemical elements and physical forces, while at the same time being the one totality everyone and everything shares, whether human or nonhuman, animate or inanimate. Without these forces, elements, and bodies, there is no atmosphere. This can only mean that atmosphere is an emergent quality that is more than the sum of its constituent parts. This "more," the atmospheric "excess" (of affect, of space, of collectivity, of surfaces—see Philippopoulos-Mihalopoulos 2015) is shared by the participating bodies, human, and nonhuman. Atmospheric emergence occurs because of this sharing—atmosphere is not an a priori waiting to be shared. Sharing, in other words, constitutes the atmosphere. Sharing is the way the atmosphere emerges. Indeed, sharing *is* atmosphere.

In the planetary atmosphere, sharing, at least apparently, only refers to substances (chemical elements and molecular structures) and processes (gravity, breathing, oxidation etc.). It does not refer to emotions. At best, it can refer to feelings about the weather or climate patterns. Yet, this distinction between material on the one hand and emotional on the other is artificial (Anderson 2014; Philippopoulos-Mihalopoulos 2015). Sharing and indeed sharing *in* an atmosphere is what Dylan Trigg (2020, 5) referring to Heidegger, calls "a matter of being-in-the-world...both singularly and plurally." But this sharing can no longer be thought to be anthropocentric. Nonhuman sensing is a technological reality (McCormack 2018). Nonhuman feeling is an ontological reality (Serres and Latour 1995). We are not the sole actors whose emotions and senses are shared with the audience below. Nor are we the only emotionally and sensorially able receptacles of material planetary processes. Both our singularity and our plurality are mediated. And as a result, *everything* feels. It is time we disengaged the emotional and the sensorial from human monopoly.

In order to do this, let's take a couple of steps back. The connection between feelings and space (and by extension, planetary space) is clear. Famously, "feelings [Gefühle] are atmospheres poured out into space" (Schmitz *et al.* 2011, 255) and "atmospheres are feelings poured out into space" (Griffero 2014, 108). I take this circularity to indicate not so much interchangeability of terms but reciprocity of flow between feelings and atmospheres. In any case, the act of pouring attempts to bridge the human body with the space around it in emotional and sensorial ways.

Schmitz's valuable contribution is that of resemiologising the classic process of *naturalisation* of the interiority of the soul and emotions that started with Plato and peaked during the European Enlightenment. Read along such thinkers as Nietzsche, Leibniz, and Whitehead, as well as feminist authors such as Rosi Braidotti, Denise Ferreira da Silva, Kathryn Yussof, and Jane Bennett, we can see how this grand project of human interiority has had some serious adverse consequences. While the dethroning of obscurantism was no doubt a necessary move away from political oppression of divinely-sanctioned social hierarchies, it has led to the ecstatic enthronement of a very specific kind of anthropocentrism that habitually naturalises or even idealises the colonial and patriarchal tendencies of its main human subject, with extreme consequences for whoever and whatever does not fit the description (Adorno and Horkheimer 2002). This naturalised individualism has posited man (and not human) as the centre of the sensorial and emotional universe. Following the successful territorial campaign of human rational thought, nonhuman emotions were quickly extinguished, and human emotions were subsumed to the same exclusionary rationality. Thus, only emotions of white heterosexual males of means and power are relevant—and these emotions were effectively castrated by the very rationality that upheld them. What remained were "female" emotions that became debased and reduced to either support act or hysteric resistance to male plans; and the non-emotions of slaves, black people, sexual deviants, disabled, animals, plants, and nature as a whole, all of which was (and to some extent still is) summarily reduced to the state of resources.

This must be read in tandem with the slow historical process, from Plato through to Descartes and Hegel, of an *engulfment*, as Denise Ferreira da Silva (2007) puts it, of the non-white, colonised subject into that vast interiority of the supreme spirit. This is the production of the "transparent I" of Western philosophy, namely the knowing human subject that is not determined by exteriority but by itself, emotionally and sensorially. These thought processes feed the fantasies of the self-sufficient, self-regulated I, as opposed to the "affectable I," the racialised other who is unable to determine itself through its own interiority (regularly succumbing to the "animality" of the sensorial) and, thus, becomes engulfed in the interiority of the universal. In turn, these racialised subjects become bestialised, serving the purpose of maintaining the exclusionary atmosphere. As Zakiyyah Iman Jackson (2020, 23) puts it, "Eurocentric humanism needs blackness as a prop in order to erect whiteness." These atmospheres of dialectically placed inclusion and exclusion are detrimental not just to the excluded (black, beast, nonhuman) but to all involved bodies: for Jackson, "Enlightenment thought [is] the violent imposition and appropriation—inclusion and recognition—of black(ened) humanity in the interest of plasticizing that very humanity" (Jackson 2020, 20).

While work by Schmitz and others working on the presocratic tradition has helped move away from the individual material body to an expanded and diffused felt body (das *Leib*) with its emotional agency (even when the conscious part is circumnavigated), the whole affair remains solidly anthropocentric (and Anthropos still defined according to the purity of the Enlightenment ideal; Grear 2015). Even if we accept that the human in atmospheric theory is not central (just always present and disproportionately decisive), there is little doubt that *he* carries on thematising the space around *him*, that surfaceless expanse as Schmitz imagines it, which, although liberated from Euclidean geometry, keeps on serving as a container for human activities. The neo-phenomenological supposed lack of control that the human has over their feelings, and consequently of the formation of atmosphere, might bring about a circular maelstrom between the (always) human subject and their object, but does not go far enough in flattening out the playing field (for my critique of phenomenology in the context of atmospheres, see Philippopoulos-Mihalopoulos 2016).

Before the naturalisation of the distinctions between subject and object, human and "nature," human and animal, feelings and senses, and indeed emotions and atmospheric phenomena, the sides were swimming alongside in the great universal penumbra, right there in the epicentre of the light-to-come promised by the classical Greek world. The Homeric and to some extent the presocratic world were characterised by a different maelstrom whose swirling sides could not be distinguished. How human were the Greek gods? How godly were the Homeric humans? Where do we place the demi-gods? Unlike the neo-phenomenological traditions, the Homeric is not an attempt at stitching back together subject and object. It is the pre-schismatic event of the continuum between humans and nonhumans (Philippopoulos-Mihalopoulos 2015): there are only epic bridges, to labour a little this metaphor, leading nowhere but to the same point. The plot (the Homeric plot but also the plot of individual lives, as we are led to understand) was unfolding while respecting the circularity of flatness. No subject versus object, inside versus outside, feelings versus senses, emotions versus planetary atmosphere. To take an example, storms and winds were not just something in which the human was enveloped emotionally (Schmitz 2007) but decisive flows that reinforced the ontological continuum amongst gods, beasts, elements, humans, half-beasts, half-gods, *and* the ineffable oral materiality of the epic texts (Purves 2010). This is more than attunement, in which a priority (of the atmospheric or at least of its conditions) is always half-assumed. Homeric works are collective emergences in which all participating bodies, human and nonhuman, move in unison towards an atmospheric totality.

The question I would like to pose here is this: how far are we today from the Homeric times, when the elements could control and, in turn, be controlled by the heroes' *nostoi* (their desire, in this case, their desire to return, to belong, to be part of something that affected the way the plot's elements would turn)? Or from when *thumos*, this great emotional and sensorial prosthesis with its own vibrancy independent of the subject, would at the same time *be* the world? Is the text of our movements and pauses so differently written to that of the Homeric texts? Are we truly the masters of our emotional fate or our sensorial preferences and tastes? How much have the advances of psychology in the 19th century and of psychoanalysis in the 20th century helped in claiming "back" our (if these were ever indeed our) emotions? Or might they have simply contributed to an atmospheric construction filled with the illusions of self-possession, freedom of choice and will power that characterise our capitalist society (Sloterdijk 2013)?

The storm is brewing, and one cannot easily tell whether it's emotions, clouds, pollution, viruses, or meteors coming our way. One realisation, however, emerges: emotions and senses are not solely human prerogatives but planetary ones. The revisiting of Homeric times is instructive, not only as a revival of the question whether the origin of emotions is inner-or outer-human, but significantly as a way of approaching a more-than-human ontological unfolding. Ben Anderson's (2009, 78) well-known formulation of what is an atmosphere is a step in the right direction: "a class of experience that occur before and alongside the formation of subjectivity, across human and non-human materialities, and in between subject and object distinctions." But it is still attached to experience and the distinction between subject and object. Gernot Böhme (2016) manages to imagine a more-than-human ontology when he describes the whole world as feelings, indeed an irradiation through space. Nevertheless, this cannot but take place before the human mirror. It seems that even in new phenomenological traditions, the human remains if not the originator, certainly the receiver of such non-human feelings. In this endeavour, even post-phenomenology (e.g., Hepach's 2021 conservative approach) that attempts to bridge new ontology with phenomenology quickly reaches its limits by failing to take its distance from the centrality of the human.

A different thought trajectory is needed that will not try to capitulate on the connection between human and nonhuman but discard the distinction altogether when it comes to the emotional sharing of atmospheres (Brown *et al.* 2019). This is what I have elsewhere (2015) called the *ontological continuum* amongst the various human and nonhuman bodies. Theories such as new materialism, object-oriented ontologies, vitalist and non-representational theories have described the end of human exceptionality, and with it, the centrality of the subject. The work of Deleuze and Guattari has been the precursor for such a move but also the connecting links to a long alternative history of philosophy that begun with the presocratics including non-western cosmogonies and moved on to Spinoza, Nietzsche, Whitehead, and so on. Deleuze and Guattari's *percepts* are instrumental, defined as "no longer perceptions; independent of a state of those who experience them." Instead, precepts are "beings whose validity lies in themselves and exceeds any lived" (1994, 164). Using literary examples, the authors show how we are no longer talking about "perception of the moor in Hardy" as we are used to "but the moor as percept; oceanic percepts in Melville; urban percepts, or those of the mirror, in Virginia Woolf. The landscape *sees*...The percept is the landscape before man, in the absence of man" (168–169). Now, couple this with Michel Serres's (Serres and Latour, 1995: 131–132) understanding of the world beyond phenomenology as affective on its own terms, generative of its own emotions, senses, and atmospheres, without relying on human consciousness, language, or even sensorial ability to bring them forth.

We inhabit neo-Homeric times, where our characters dissolve in the grand atmosphere, becoming "part of the compound of sensations" (Deleuze and Guattari 1994, 169). What could, however, be a space of exploration, with respect for difference and an understanding of how collective bodies operate, or what Dylan Trigg (2020, 1) calls "a common ground between people, in turn generating a framework for mutual self-other awareness," it is simply an overarching atmosphere directing affects in asphyxiatingly precise ways. The planet has been reduced to a subaltern to the human, and, in turn, the human reduced to a subaltern to a global atmospheric engineering. Our neo-Homeric gods and elements are no longer Olympian but crushingly quotidian. Every process, product, sense, thought has been put in the service of such atmospheric engineering. This engineering that pushes humanity towards fixed ideas of progress and growth while abusing vast parts of animate and inanimate bodies is so tight and precise, so all-involving that there is hardly any manoeuvring space except for the participating bodies, human and nonhuman, to obey whenever interpellated by the atmosphere. This is the era of the Anthropocene, characterised at the same time by a human colonising omnipresence and an abject human absence. We are characters in a plot that seems stuck. And there is yet another difference to the Homeric times: we are in it because we think we have the freedom to choose whether to be in it or not. Our freedom, however, is conditioned by our very own desire not to be free. The Homeric maelstrom has been replaced by a frightening circularity: we are all pawns of vast atmospheric processes; yet these atmospheric processes rely on our desire to be part of them. Our *nostos* is now narrowly defined as the panicky return to the immunisation of atmospheres. Our *thumos* is coagulated into a monodimensional self-feeding desire for more of everything, as Böhme has it: "desires cannot be permanently satisfied, but only temporarily appeased, since they are actually intensified by being fulfilled" (Böhme 2016, 11).

Atmospheres determine our (false) desire. Our (false) desire determines the atmosphere. Everything is now placed in the service of inescapability.

And in the meantime, our planet is being destroyed, animals and plants are treated as infinitely renewable resources, and large swaths of the global population are being exploited.

## Sharing is caring

The above are political necessities, presenting themselves with urgency in view of the ecological crisis, the systematic marginalisation of "othered" bodies, and the demotion of justice to an after-thought when compared to profit and individual interest. In a fast-changing ethicopolitical sphere where even fossil fuels are being divested of funding, animal rights are widely understood to be the minimum guarantees for animal welfare, and issues such as Black Lives Matter are having such an impact on the everyday, atmospheric research risks becoming complicit with the status quo. We can certainly carry on observing, assessing, and theorising about atmospheres. But unless we assume the ethical and political responsibility of our research, our texts risk losing both relevance and vigour.

The question, therefore, is what can be done about this through atmospherology and the question of sharing. In its expansive understanding, sharing (of emotions, meteors, and nonhuman precepts) is an important practice: Barbara Koziak (1999) in her feminist take on Homeric emotions shows how Achilles and Priam share a powerful (and "unmasculine") emotion, that of sorrow, which is co-emergent with the inauguration of the atmospherics of a new polis. This is close to Trigg's sharing formulation "If shared emotion reinstates and reinforces an overarching atmosphere, then we are not only attuned to an atmosphere in a passive and structural sense; we are also attuned to an atmosphere insofar as it gives expressive form to a set of evaluative concerns that serve as a basis for integrative togetherness" (2020, 5). Drawing on Merleau-Ponty's operative intentionality, Trigg offers an affective sharing of an atmosphere "both diffused in the air and grasped under the skin" (2020, 4). While Trigg's take remains anthropocentric and focused on the animate, I would like to imagine that this air found both outside and inside enables a posthuman integrative togetherness.

One needs to tread carefully though. Elements, such as air, hold the promise of connection and liberation while at the same time becoming readily conditioned for oppressive atmospheric purposes. Air, for example, is a central part of the elemental basis of atmospheres (McCormack 2018; Adey 2010) because it is found both inside and outside the animate and the inanimate, permeating everything and imparting life and death through breath and oxidisation. For this reason, air is regularly staged, conditioned, manipulated, and engineered (Philippopoulos-Mihalopoulos 2016). Air conditioning is not just a metaphor but the elemental terrain beyond human consciousness where atmospheres operate. Marie-Eve Morin (2009, 68) writes: "as our management and regulation of the atmosphere is now part of our social and political concerns, the atmosphere can no longer be seen as an exterior but has effectively been integrated within the system of human relations." Timothy Choy (2011) talks about how atmospheric governmentality amounts to a suspension of ordinary forms of experience, with obvious ecological repercussions in terms of concealed differences of atmospheric pollution measurements across cultures. Choy calls attention to air's relative under-theorisation in relation to solidity (earth, water) and draws specific conclusions on spatial inequality of income and human atmospheric permeability that point directly to a causal link between occlusion of air from analysis and atmospheric manipulation.

For an integrative togetherness, for a true sharing as caring ontology, human emotions are not enough. True integrative togetherness can only occur when the planetary is given its due, and when sharing is truly material and inclusive. Atmospheric theory has an important role to play, but this role might sound counterintuitive: the best way to endeavour for an integrative sharing is by *withdrawing* from atmospherics. As I have shown elsewhere (2015), even well-intentioned atmospheric engineering risks serving manipulating purposes. Even when it aims at calm, comfort, *hygge* or *Gemütlichkeit*, an atmosphere can be put in the service of happiness surveys, productivity incentives, heteronormative proliferation of social models, and so on (Anderson 2014). Böhme's work (e.g., 1995, 2013, 2016) shows how we are regularly subsumed to the way atmospheres are aesthetically staged. We are bombarded from every single angle. Lauren Berlant (2011) shows how the publics as affective machines of persuasion have thematised American society through popular films. Zizi Papacharissi (2015) shows how social media allow for the production of affective textures with an affective quality, which, in turn, attract constantly new audiences. Finally, Sloterdijk (2013, 198) describes atmospheres as "a climatized luxury shell in which there would be an eternal spring of consensus." This is not a new phenomenon, but social media and the subsequent extreme social and political polarisations as we have seen with, e.g., Trump's politics via tweets, Covid-19 and the mask polarisation, or the various conspiracy theories about global political power, vaccinations or indeed the existence of Covid-19 itself, have intensified and accelerated such atmospheric persuasion. At the same time, we must understand the role our desire plays in atmospheres (this evanescent but presumed consensus), and how, in turn, this reinforces the power of atmospheres to direct our desire (towards more atmospheric consensus). We must understand and break this circularity by withdrawing from atmospherics. This presupposes two things: first, that we know how atmospheres are engineered and how this engineering affects us and the planet; and, second, that withdrawal from atmospherics must take place in a climate of understanding and forgiveness—or, and, atmospherics, even in their engineered form, are ways in which bodies organise themselves in order to deal with the vast openness of life.

## Atmospheric conditions

Along the lines of Peter Zumthor's (2006) atmospheric instructions, I offer here four conditions that must be met for an atmosphere to be successfully engineered. But my intention could not be more different from that of Zumthor's. Indeed, here I caution against atmospherics by showing how they employ our desire circularly, making us desire-addicts. I do this in full avowal of my own atmospheric desire, my own comfort zone barricaded behind notions of belonging while simultaneously forgetting the violence of engulfment that inevitably takes place.

1   *An atmosphere must generate a distinction between interior and exterior.* Atmospherics imply the violence of partitioning inside from outside. In view of the ontological continuum, any distinction between inside and outside is artificial. Yet this is the main atmospheric technology. While the practice of partitioning, enclosing, or enveloping is not always accepted or adequately emphasised in the literature, I find that not doing so is a misrepresentation of the political violence that is always resident in any atmosphere, however benign. There is a very simple explanation: in order for an atmosphere to emerge, some bodies must be retained while other bodies must be excluded. Think of the planetary atmosphere and the way it works in parallel with gravity in order to retain the very elements of atmosphere while excluding any intruder. This distinction between retaining and excluding, or inside and outside, is the most fundamental notion in sociologist Niklas Luhmann's systems theory (2012), who has significantly influenced Sloterdijk's work, and which can be of use here too. In Luhmann's systems theory, the (social) system builds its identity by excluding its exterior (whatever the system is *not*). Once inside the systemic interior, one cannot see outside, "atmospheres are made available as total settings of attractions, signs and contact opportunities" (Sloterdijk 2004, 180). This gesture finds its expression in atmospherics: partitions can be physical (gated communities), semi-physical (e.g., orthodox Jewish *eruv*), or entirely notional (our comfort zone, supermarket packaged meat, citizenship, and so on). The compartmentalising lines are arbitrary, drawn post-facto, fed by the desire to return to the safe atmospheric embrace (and the subsequent desire to extend the reach of that embrace so that all possible future threats are pre-emptively eliminated), which might have little to do with actual risk. Senses, emotions, and symbols are orchestrated in such a way that the partition is reinforced and never questioned (Borch 2014; Closs 2015). The West is a glasshouse of atmospheric partitioning, with immigration policies controlling the use of elements such as water and land in terms of spatial approaches to sovereign utopias, or the boundary that separates the occident from the orient, constructing both exteriors and interiors through religion, economy, culture

and so on. As Frantz Fanon (1963, 37) writes, "the colonial world is a world divided into compartments." Racial violence has often been at the core of atmospheric engineering in the form of racial threat (when in white atmospheres) or racial discrimination and oppression (when in non-white atmospheres). Tayyab Mahmud's (2010) work on postcolonial spaces of oppression shows this amply. Slums are atmospheric constructions where "surplus humanity" is piled up and kept inside through atmospheric techniques of accumulation through dispossession and primitive accumulation (namely Marx's concept of deprivation of the means of subsistence). These techniques define the exterior of the slum as a non-possibility, thus strengthening what can be described as negative belonging, that is, belonging because of the impossibility of belonging anywhere else.

Perhaps the most important aspect of this manipulation, however, is our complicity with it. Indeed, the great unifier of the interior is the desire of individual bodies to be part of the atmosphere—for as we shall see, even an engineered atmosphere must never feel imposed but always emerge from the desire of the participating bodies. What we forget of course is that the distinction between inside and outside is entirely artificial, as we have seen with the locus of emotions. But perhaps the greatest tragedy for humans would be the absence of an inside. So we build vast exclusionary structures and cleaving conceptual contrivances, just to be able to shut the door behind us.

2  *The exterior must be included in the interior.* Or to put it simply, the exterior becomes a resource for the perpetuation of the atmospheric interior. The outside is not actually excluded but drained and manipulated into the position of the subaltern. Think of how nature is partitioned from the human and then considered a resource without intrinsic value except in relation to human needs. Or how slavery was justified on the basis of civilisation's needs. This is as we have seen is what Ferreira da Silva (2007) calls *engulfment*. The result is that the interior becomes so advanced that it needs contrast, diversion, support, hinterlands, colonies, slaves, animal farms, felled tropical forests, a sacrificial planet: "a cyclic or a spherical shape with an inner limit and an outer boundlessness" (Negarestani, 2008: 102). In most cases, the predominant sense within an engineered atmosphere is one of contentment: we have all we need right here. An engineered atmosphere is "an enclosure so spacious that one would never have to leave it" (Sloterdijk, 2013, 175), with oodles of supposed freedom at our disposal. Or that one *could* never leave: the diametrically opposite affect emerges in atmospheres of oppression and legal hyper-visibilisation, such as kettling police strategies, prisons, or court rooms. Atmospheres of oppression rely on a constant redrawing of spatiotemporal boundaries: Vaughan-Williams talks about the way mobile phone operators can be ordered to close their network, as it happened in the 22/7 London bombings. While this was in order to help the

police carry out its duties, it also denotes a new spatiotemporal zoning power from which nobody can escape: "such decisions are no longer localized or fixed at particular border sites in the margins of sovereign territory but increasingly more widespread or diffused throughout society: a phenomenon that might be captured by the concept of a biopolitical generalized border" (Vaughan-Williams 2007, 191). The greater the inclusion of the exterior, the greater the (impression of) risk in the interior. In his oeuvre on community, Roberto Esposito (2011, 141) writes: "as in all areas of contemporary social systems, neurotically haunted by a continuously growing need for security, this means that the risk from which the protection is meant to defend is actually created by the protection itself." For this reason, an atmosphere includes and pacifies its own risks: successful atmospheres include and anticipate even doubts, second-thoughts, objections, resistance, or revolts against it. A true totality building on the desire of the participating bodies to perpetuate these distinctions.

3  *An illusion of synthesis is offered inside every atmosphere.* For Sloterdijk (1998), a sphere is the shared circle of humanity. The fact that humans dwell in a sphere *makes* them human—or this is what atmospheric rhetorics wants us to believe. The illusion of synthesis as a *human* sphere is one of the triumphs of the atmospherics of humanism. The membrane around the sphere protects the interior from the exterior and offers Sloterdijk's much-valued immunity. Feeling safe is a collective affect that ontologically translates into the following: any continuum needs to be ruptured in order to be tolerable. Likewise, with feeling part of a "disrupting" or "protesting" crowd: to be inside is to be safely identifying with "the right side." Comfort is no more impressionistic in a kettled enclosure than it is in a shopping mall. To talk about comfort in kettling sounds odd, but comfort comes from a desire to belong, especially when the delineation is unambiguous. One needs to fold the continuum around oneself in order to *feel* (comfortable, safe, belonging, present, alive, vindicated, heroic): this fold ruptures the continuum. This means that in order for the atmosphere to be perpetuated, it needs to be contained as an isolated instance of the continuum, as a rupture which refines the continuum according to the specifics of the atmosphere. For we must not forget that atmospheres are fractal emergences, one incubated in the other, and all incubated in the one total global atmosphere.

In an atmosphere, the synthesis finds its apogee in questions of time. The future is frozen or at least perfectly predictable (and safe). Atmosphere resides in an eternal present. Nothing will change. Nothing has to change. The inherent conservatism of atmospheres is dissimulated as a reassuring promise of stability, presence, and continuation. Is this a shopping mall or a prison? We are conditioned by the air inside, a round present vibrating with vapid promises.

4  *An atmosphere must dissimulate the fact that it is engineered.*
Dissimulation means: no one has engineered the atmosphere, no one
has organised matter and non-matter to generate the atmosphere.
Atmosphere dissimulates its engineering and presents itself *only* as an
emergence. This is why it is in the interest of an atmosphere to carry on
being described as evanescent, elusive, uncapturable, quasi-, or half-
entity. The vaguer, the better! But really, what's mysterious about political manipulation? What's elusive about the oldest rule in the book,
divide and conquer? What's intangible about the way people react to
manipulation? A quick foray into supermarket marketing shows that
it is an exact science. The fact that engineered atmospheres are *often*
but not *always* successful, does not make their engineering unscien-
tific, mysterious or occult. Percentages are well within the scientific
turf. If enough people are persuaded to buy the specific toothpaste,
marketing science knows that their advertising worked. Or if the lighting, movement training, and colour co-ordination works for the bright
young things who live the life of voluntarily permanent Big Brother
objects in huge LA villas and produce photogenic yet extremely controlled "content" for their constantly expanding Instagram followers,
then the product they'll endorse will simply be bought as many times
as they expect. Nothing ineffable in this. Or perhaps the need to maintain the fantasy of atmospheres as ineffable is simply a reflection of the
burning desire to assert our own individualism: we are all different! No
one can predict how each one of us will react. We all bring something
different to the atmosphere, and the atmosphere changes according
to that. But are we really? Is there really such a difference amongst
us? Are the way we react to the powerful attraction of atmospherics
so unpredictable? Do we really manage to disrupt them and diversify
them enough? The only ineffable thing is the strength of this wildly
individualising fantasy. And here is the grand atmospheric trick:
atmospheric self-dissimulation means that an atmosphere dissimulates
itself (as well as its origin and its nature) as non-atmosphere, as mere
habit, practice, historical fact, cultural norm, temporal necessity, and
indeed *desire*, where there is always a possibility of non-scripted imbal-
ances, excesses, conflicts, or revolts to take place. During that dissimulation, old binarisms remain but change their nomenclature: thus from
"Civilised/Barbarian, Believer/Infidel, White/Black or Advanced/
Primitive" to the more "acceptable" "Developed/Developing, Centre/
Periphery, Advanced/Emerging, or Rich/Poor" (Eslava and Pahuja
2012, 2). In a self-dissimulating atmosphere, that most accomplished
of atmospheres, there is nothing to go against: the atmosphere has converted itself into Quixotic windmills. Not only are we all made to be
complicit with the atmosphere and its exclusions, but we are also made
to feel guilty for desiring its perpetuation.

**Where do we withdraw from here?**

The onus on the individual or the collective to tell the difference between (real) desire and (false) desire is perhaps the bane of our era (Braidotti 2013). The aesthetic economy (Böhme 2016) of atmospheres means that everything banks on its attractiveness in order to wake up our desire for it. Desire becomes perceived as need and need translates into radical atmospheric engineering.

Withdrawing from atmospheres is not a passive move, nor apathy or indifference. It is, first, based on the idea that one attempts to tell the difference between desire and desire. We have reached the point where we do not desire something because it is good for us. Rather we judge something to be good because we desire it. Telling the difference between desire and desire (joy and sadness, in Spinoza's terminology of affect) is a question of knowing the *causes* of something: why it is that we are desiring what we are desiring? How good is it for us in relation to the assemblage to which we belong? To quote Spinoza (1992, 286, translation modified, emphasis added), "this, then, is that human freedom which all humans boast of possessing, and which consists solely in this, *that humans are conscious of their desire and unaware of the causes by which they are determined.*" Second, and related to the above, atmospheric withdrawal presupposed that one becomes aware of the excluding and engulfing effect that one's atmospheric participation has on other bodies and the planet. And, third, that one does not act alone in this but is always part of a collectivity, an assemblage whose desire is directed away from consumerist, unethical, and exploitative "choices" and towards true acts of alternative existence. But one individual is never enough. We have run out of Hollywood heroes. It might have worked in *Truman Show* where, when Truman apprehends the atmosphere of enclosure and perfection and tries to *withdraw*, while the voice of his "creator," the eternal Father, god, atmosphere itself—namely the reality show director—booms from high up the set dome, simultaneously reinstating and allowing the atmosphere to become one of terror. That's nice and cinematographically pleasing, but ultimately simply another atmospheric construction that promotes heroism as essentially an individual trait of a single leader. We need to share: we need to ride a wave of collective change amongst other bodies, human and non-human, in order to withdraw from the powerful attraction of atmospherics. So *sharing coming full circle, as a way out of atmospherics.*

"The revolutionary knows that escape is revolutionary—*withdrawal, freaks*—provided one sweeps away the social cover on leaving, or causes a piece of the system to get lost in the shuffle. What matters is to break through the wall" (Deleuze and Guattari 1983, 277). Desire captures, desire allows to withdraw. One withdraws by acting on desire, while at the same time acting *against* desire. *Withdrawal is self-withdrawal*: a body withdraws from the space of its own desire, the one that keeps the body atmospherically conditioned. But since it is always the bodies themselves that constitute an atmosphere, to withdraw from an atmosphere is always an act of self-withdrawal. An atmosphere is inscribed on the body in the same way as the body inscribes the atmosphere: through desire. To withdraw is to counter-pose one's body against the body of one's desire.

So what happens when we withdraw? Maybe nothing. We may just land on another atmosphere. But the movement might be enough to shake things up, indeed to transform things. A transformation that can only happen through reorientation. Jean-Francois Lyotard (1988, 9) has put it memorably with his concept of the *differend*: "A case of *differend* between two parties takes place when the 'regulation' of the conflict is done in the idiom of the one of the parties, while the wrong suffered by the other is not signified in that idiom." Withdrawing from the atmospheric plenitude of wilful oppression, biopolitical control, spatial partitioning, and symbolic/linguistic propaganda aims at a new game, indeed a new register, with new rules and new hopes. Caution though: all new games tend to coagulate into atmospheres. As soon as the rules are repeated and the exclusions become established, hope is frozen into co-opted atmospheric business-as-usual.

We need to keep on moving.

## Conclusion

Allow me to close with a reminder: while withdrawing is necessary, atmospherics is also necessary. And we must be gentle, both to ourselves and others when we indulge in our desire for atmospherics, even when such desire turns noxious, soul-eating, life-destroying.

I was fortunate to see an artwork by Ingeborg Lüscher at Hamburger Bahnhof in Berlin a few years ago called *The Other Side: Israel/Palestine*. The artwork consisted of three long horizontal screens situated next to each other. In a series of silent black-and-white relatively brief takes, the faces of approximately thirty Palestinians and Israelis were shown, one at a time. The face appeared on the first screen, only to disappear afterwards and reappear on the second screen and then finally the third. Every time, how-ever, the expression was different, as if something had happened to which we were not privy. I then noticed that three plaques were positioned under-neath each screen. The first read: "Think. Who are you, your name, your origin?" The second: "Think. What has the other side done to you?" All the participants were asked the same questions. All of them had lost loved ones during the conflicts. Lüscher was filming them while asking these ques-tions, which we could not hear. We were not told who was on which "side," which atmosphere they dwelled in, and, although one could guess, the takes were meant to conflate them rather than to keep them as "sides." The varia-tion and emotional impact of the facial expressions of the participants were overwhelming. The usual trajectory was one of pride and defiance mixed with pain; this would then change on the second screen to intense pain and increasingly deepening sorrow. But the most devastating moment was when the third screen would come alive with the participant's face. It was the screen that betrayed most expectations and went against most projections of how the participants would react and how their expressions would change. These last screens were a humble triumph against synthesis. They could not be predicted on the basis of the previous screens. They emerged from a withdrawal from the atmospherics of conflict so powerful that it was humbling, so defiant that it was devastating.

The plaque underneath the third screen read "Think. Can you forgive?"

## References

Adey, Peter. 2010. *Aerial Life*. Malden, Mass: Wiley-Blackwell.

Adorno, Theodor and Max Horkheimer. 2002. *Dialectic of Enlightenment*. Trans. E. Jephcott. Stanford: Stanford University Press.

Ahmed, Sarah. 2006. *The Cultural Politics of Emotion*. Edinburgh: Edinburgh University Press.

Anderson, Ben. 2009. "Affective atmospheres". *Emotion, Space and Society* 2: 77–81.

Anderson, Ben. 2014. *Encountering Affect: Capacities, Apparatuses, Conditions*. Farnham: Ashgate.

Berlant, Laurent. 2011. *Cruel Optimism*. Durham: Duke University Press.

Bille, Mikkel, Peter Bjerregaard, and Tim Flohr Sørensen. 2015. "Staging atmospheres: Materiality, culture, and the texture of the in-between." *Emotion, Space and Society* 15(1): 31–38.

Bille, Mikkel. 2015. "Lighting up cosy atmospheres in Denmark." *Emotion, Space and Society* 15(1): 56–63.

Böhme, Gernot. 1995. *Atmosphäre: Essays zur neuen Ästhetik*. Suhrkamp, Frankfurt am Main.

Böhme, Gernot. 2013. "The art of the stage set as a paradigm for an aesthetics of atmospheres." *Ambiances International Journal Sens. Environ. Archit. Urban Space.* http://ambiances.revues.org/315

Böhme, Gernot. 2016. *Critique of Aesthetic Capitalism*. Trans. Edmund Jephcott. Rome: Mimesis International.

Borch, Christian, 2014. "The Politics of Atmospheres; Architecture, Power, and the Senses." In *Architectural Atmospheres: On the Experience and Politics of Architecture*, edited by Christian Borch. Basel: Birkhäuser.

Braidotti, Rosi. 2013. *The Posthuman*. Cambridge: Polity.

Brown, Steven D., Ava Kanyeredzi, Laura McGrath, Paula Reavey, and Ian Tucker. 2019. "Affect theory and the concept of atmosphere." *Distinktion: Journal of Social Theory* 20(1): 5–24.

Choy, Tim. 2011. *Ecologies of Comparison: An Ethnography of Endangerment in Hong Kong*. Durham: Duke University Press.

Closs Stephens, Angharad. 2015. "The affective atmospheres of nationalism". *Cultural Geographies* 23 (2): 181–198. doi:10.1177/1474474015569994.

Deleuze, Gilles, and Félix Guattari 1983. *Anti-Oedipus*. Trans. R. Hurley, M. Seem, and H. R. Lane, Minneapolis: University of Minnesota Press.

Deleuze, Gilles, and Félix Guattari. 1994. *What is Philosophy?* Trans. H. Tomlinson and G. Burchell. New York: Columbia University Press.

Eslava, Luis, and Sundhya Pahuja 2012. "Beyond The (Post)Colonial: Twail and The Everyday Life of International Law." *Journal of Law and Politics in Africa, Asia and Latin America -Verfassung und Recht in Übersee (VRÜ)* 45(2): 195–221.

Esposito, Roberto. 2011. *Immunitas: The Protection and Negation of Life.* Trans. Z. Hanafi. Cambridge: Polity Press.

Fanon, Frantz, 1963. *The Wretched of the Earth.* Trans. C. Farrington. New York: Grove Press.

Ferreira da Silva, Denise. 2007. *Toward a Global Idea of Race.* Minneapolis: University of Minnesota Press.

Grear, Anna. 2015. "Deconstructing Anthropos: A critical legal reflection on 'Anthropocentric' Law And Anthropocene 'Humanity'." *Law and Critique* 26 (3): 225–249.

Griffero, Tonino. 2014. *Atmospheres: Aesthetics of Emotional Spaces.* Trans. Sarah De Sanctis. Farnham: Ashgate.

Hepach, Maximilian Gregor. 2021. "Entangled phenomenologies: Reassessing (post) phenomenology's promise for human geography," *Progress in Human Geography* 1–17. DOI: 10.1177/0309132520987308

Hines, Charles. 1972. "Gravity waves in the atmosphere." *Nature* 239: 73–78.

Jackson, Zakiyyah Iman. 2020. *Becoming Human: Matter and Meaning in an Antiblack World.* New York: NYU Press.

Koziak, Barbara. 1999. "Homeric *Thumos*: The Early History of Gender, Emotion, and Politics." *The Journal of Politics* 61, no. 4 (November): 1068–1091.

Luhmann, Niklas. 2012. *Theory of Society, Volume 1.* Trans. R. Barrett. Stanford: Stanford University Press.

Lyotard, Jean-François. 1988. *The Differend: Phrases in Dispute.* Trans. G. van den Abbeele. Minneapolis: University of Minnesota Press.

Mahmud, Tayyab. 2010. "'Surplus Humanity' and margins of law: Slums, slumdogs, and accumulation by dispossession." *Chapman Law Review* 14(1): 10–26.

Morin, Marie-Eve. 2009. "Cohabitating in the globalised world: Peter Sloterdijk's global foams and Bruno Latour's cosmopolitics." *Environment and Planning D: Society and Space* 27: 58–72.

McCormack, Derek. 2018. *Atmospheric Things.* Durham: Duke University Press.

Negarestani, Reza. 2008. *Cyclonopedia: Complicity with Anonymous Materials.* Melbourne: re-press.

Nietzsche, Friedrich. 2005. *Thus Spoke Zarathustra.* Trans. G. Parkes. Oxford: Oxford University Press.

Papacharissi, Zizi. 2015. *Affective Publics: Sentiment, Technology, and Politics.* Oxford: Oxford University Press.

Philippopoulos-Mihalopoulos, Andreas. 2015. *Spatial Justice: Body Lawscape Atmosphere.* London: Routledge.

Philippopoulos-Mihalopoulos, Andreas. 2016. "Withdrawing from Atmosphere: An Ontology of Air partitioning and Affective Engineering." *Environment and Planning D: Society and Space* 34(1), 150–167.

Purves, Alex. 2010. "Wind and Time in Homeric Epic." *Transactions of the American Philological Association* 140(2 Autumn): 323–350.

Schmitz, Hermann. 2007. *Der Leib, der Raum und die Gefühle.* Bielefeld, Locarno: Sirius.

Schmitz, Hermann, Rudolf Owen Müllan, and Jan Slaby. 2011. "Emotions outside the box—The new phenomenology of feeling and corporeality." *Phenomenology and the Cognitive Sciences* 10: 241–259.

Serres, Michel, and Bruno Latour. 1995. *Conversations on science, culture and time.* Trans. R. Lapidus. Ann Arbor: University of Michigan Press.

Sloterdijk, Peter. 1998. *Sphären I: Blasen: Mikrosphärologie.* Frankfurt am Main: Suhrkamp.

Sloterdijk, Peter. 2004. *Sphären III: Schäume: Plurale Sphärologie.* Frankfurt am Main: Suhrkamp.

Sloterdijk, Peter. 2013. *In the World Interior of Capital: Towards a Philosophical Theory of Globalization.* Cambridge: Polity Press.

Spinoza, Baruch. 1992. *Ethics, Treatise on the Emendation of the Intellect and Selected Letters.* Trans. E. Shirley. New York: Hackett Publishing.

Thrift, Nigel. 2007. *Non-Representational Theory: Space | Politics | Affect.* London: Routledge.

Trigg, Dylan. 2020. "The Role of Atmosphere in Shared Emotion." *Emotion, Space and Society* 35: 100658. doi:10.1016/j.emospa.2020.100658.

Vaughan-Williams, Nick. 2007. "The shooting of Jean Charles De Menezes: New border politics?" *Alternatives: Global, Local, Political* 32 (2): 177–195. doi:10. 1177/030437540703200202.

Zumthor, Peter. 2006. *Atmospheres.* Basel: Birkhäuser.

# Index

Printed in the United States
by Baker & Taylor Publisher Services